Boilers
Performance & Testing

by D. James Benton

Copyright © 2021 by D. James Benton, all rights reserved.

Preface

Boilers consume fossil fuels (coal, fuel oil, or natural gas) and produce steam for industrial processes or power production. Boilers have been a mainstay since the dawn of the Industrial Revolution. In the 21st century boilers are being replaced with green alternatives and renewable resources. The transition will take years. While this change is implemented, it is imperative that the remaining boilers be operated efficiently, which requires clear expectations and diligent testing. This text describes what can be achieved and how to verify. Examples are given in both English (U.S. Customary) and SI units.

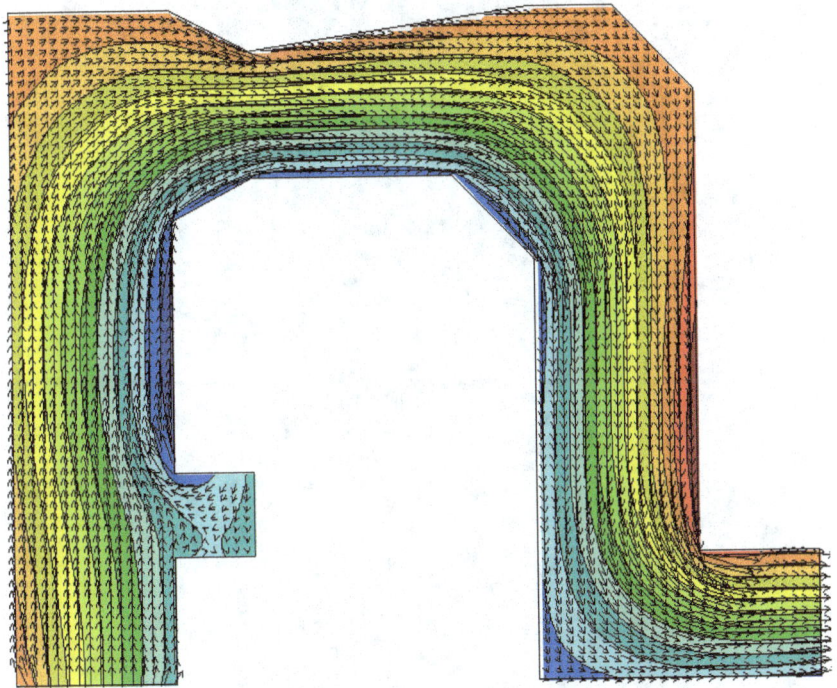

Figure 1. Idealized Flow through a Boiler

All of the examples contained in this book,
(as well as a lot of free programs) are available at...
https://www.dudleybenton.altervista.org/software/index.html

Figure 2. Typical Coal-Fired Boiler with Supporting Equipment

Table of Contents

	page
Preface	i
Chapter 1. Introduction	1
Chapter 2. Combustion	3
Chapter 3. Heat Transfer	10
Chapter 4. Uncertainty	16
Chapter 5. Gas and Oil Firing	23
Chapter 6. Incomplete Combustion	30
Chapter 7. Mass Leakage	37
Chapter 8. Sensitivity	44
Chapter 9. Solutions	49
Chapter 10. Correction Curves	56
Chapter 11. Testing	67
Chapter 12. Chemical Reactions	75
Appendix A. Higher vs. Lower Heating Value	79
Appendix B. Coal Heating Values	80
Appendix C. Natural Gas Heating Values	82
Appendix D. Flue Gas Properties	83
Appendix E. Moist Air Properties	85
Appendix F. Steam Properties	88
Appendix G. Brent's Method	89
Appendix H. Units	91
Appendix I: Modified Broyden's Method	92

Figure 3. Typical Water Wall

Chapter 1. Introduction

Boilers have two primary aspects: 1) combustion and 2) heat transfer. There are countless industrial processes that involve many types of combustion, but boilers represent only a limited subset of this whole. There are also myriad industrial processes that involve heat transfer and numerous designs for heat exchangers, but boilers are very much the same. Boilers serve a single focused purpose: burn fuel to create steam.

Direct contact between the combustion and heat transfer elements is the most effective way of achieving this end. In a boiler the fuel are mixed and burned inside the core, which is surrounded by and consists of tubes and/or sheets to directly transfer heat to the water for the production of steam. Tubes are often welded side-by-side to form a "water wall" as illustrated on the previous page. This conserves material and also eliminates any non-essential pathways for the heat to traverse when moving from the fire to the water.

As combustion takes place inside the boiler, deposits, fouling, and slag are unavoidable so that any effective design must consider these, providing resilience and also maintainability. Elaborate support systems (for example, soot blowers) have been developed to facilitate cleaning, which must be done periodically. This is a necessary cost of operating a boiler.

Combustion and heat transfer often takes place in separate devices other than boilers so that these two processes may be considered separately. When analyzing a boiler, these must be considered together. Efficiency is one performance measure that is often evaluated for each device and/or process individually. For a boiler, efficiency is an integral measure of the whole.

While there are some systems that have separate heat recovery systems that are not an integral part of the boiler, we will not consider those here. We will only consider integral parts, such as preheaters and superheaters. Scrubbers, which may be attached to a coal-fired boiler to remove SO_2 are a completely separate device and are not considered part of the boiler.

All of the calculations described in this text are implemented in the form of Excel™ spreadsheets, which are available at the link beneath the Preface.

A Heat Recovery Steam Generator (HRSG) is a special type of heat exchanger attached to the exhaust of a combustion gas turbine for the purpose of extracting energy to create steam and ultimately electricity. These are covered in a separate text, also available from Amazon.

The different sections of a boiler are strategically positioned to take best advantage of the available heat and natural temperature distribution along the path from the combustion to the exhaust or flue. These sections have separate tube bundles, which may or may not extend into the flow, depending on the section and specifics of the design. Below the evaporator, superheater, and economizer are all steam heat exchange zones, while the air pre-heater is a gas-to-gas heat exchange zone.

We know from the 2nd Law of Thermodynamics that greater temperature differences mean more entropy generated, but temperature difference is what drives heat transfer and controls required surface area and ultimately material expense. The juxtaposition of these two competing principles inspires finesse in boiler designs, which draws upon many decades of experience.

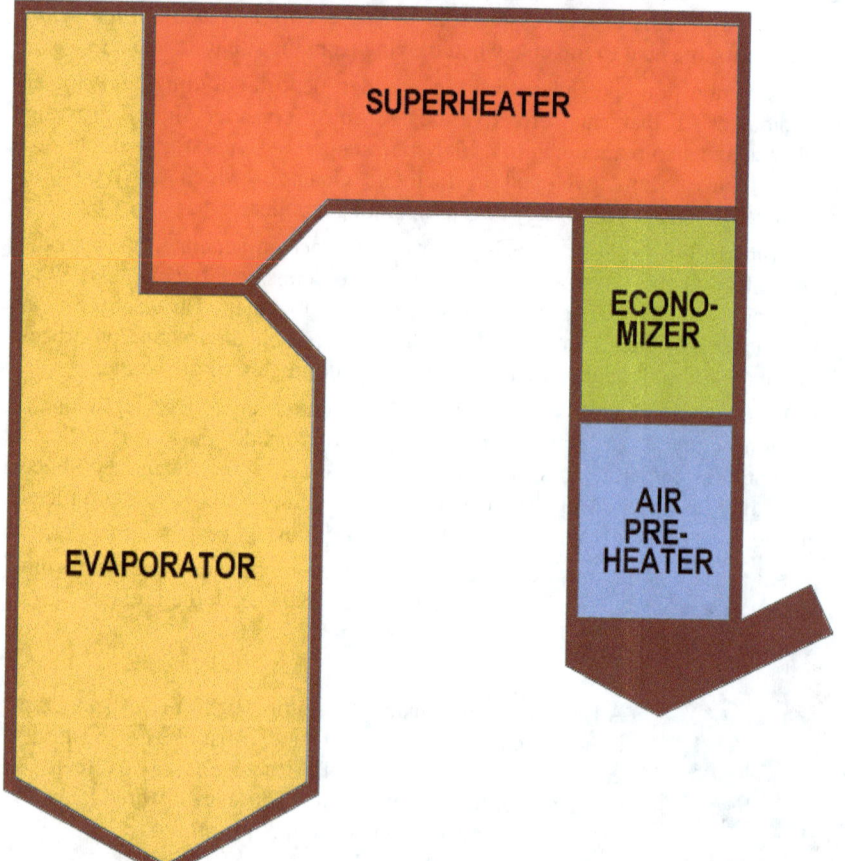

Figure 4. Basic Sections of a Typical Coal-Fired Boiler

Chapter 2. Combustion

In this chapter we consider applied chemistry as it relates to boilers. We will cover the combustion of three main fuels: 1) coal, 2) fuel oil, and 3) natural gas. Any other fuel will follow these same principles. In order to introduce this topic we must first define a number of symbols and formulas.

It is customary to use the symbol, x, for mole fractions and the symbol, y, for mass fractions; however, we will be discussing dry air, moist air, dry coal, wet coal, coal with ash, coal without ash, and flue gas. These symbols will already have subscripts for carbon, hydrogen, oxygen, sulfur, and more. It is not practical to designate all of these qualifiers as subscripts (e.g., $x_{moist.air.N2}$ for the nitrogen mole fraction of moist air). Instead, we will use other symbols; for example r for wet coal mass fractions with ash, which is how a laboratory most often reports the analysis.

Coal Composition

Coal composition (wet including ash) can be defined by mass ratios and are given the symbol, r:

$$r_C C + r_H H + r_S S + r_O O + r_N N + r_{Ash} Ash + r_{H_2O} H_2O \quad (2.1)$$

We assume that the ash does not participate in the combustion reaction, so we eliminate it and adjust the mass fractions accordingly. It is also convenient to separate the moisture (H_2O) into hydrogen and oxygen. The resulting mass fractions are given the symbol, x:

$$x_C C + x_H H + x_S S + x_O O + x_N N \quad (2.2)$$

The x components can be calculated from the r components as follows:

$$x_C = \frac{r_C}{1 - r_{Ash}} \quad (2.3)$$

$$x_H = \frac{r_H + 2 r_{H_2O}}{1 - r_{Ash}} \quad (2.4)$$

$$x_S = \frac{r_S}{1 - r_{Ash}} \quad (2.5)$$

$$x_O = \frac{r_O + r_{H_2O}}{1 - r_{Ash}} \quad (2.6)$$

$$x_N = \frac{r_N}{1 - r_{Ash}} \quad (2.7)$$

The gross (with ash) and net (without ash) coal flow rates are:

$$\dot{m}_{gross} = \dot{m}_{net} + \dot{m}_{Ash} \qquad (2.8)$$

$$\frac{\dot{m}_{net}}{\dot{m}_{gross}} = \frac{\dot{m}_{gross} - \dot{m}_{Ash}}{\dot{m}_{gross}} = 1 - r_{Ash} \qquad (2.9)$$

Composition of this particular coal on a mass and mole basis is shown in the following table:

Table 1. Composition of Coal (with Ash)

ox. num.	mol. wt.	by mass	by mole	
1.00	12.011	48.51%	41.141%	C
0.25	1.008	3.25%	32.843%	H
-0.50	15.999	10.69%	6.806%	O
0.00	14.007	0.65%	0.473%	N
1.00	32.060	0.40%	0.127%	S
0.00	51.746	5.50%	1.083%	Ash
0.00	18.015	31.00%	17.528%	H_2O
0.46075	10.186	100.00%	100.00%	total

The effective molecular weight is 10.186 (shown at the bottom). The oxidation numbers will be used later. The first (1.00) means that combustion of one mole of carbon (C) requires one mole of oxygen (O_2). The second (0.25) means that combustion of one mole of hydrogen (H) requires one-half mole of oxygen (O_2) to form one-half mole of water (H_2O).

Air Composition

While dry air does vary from one location and elevation to another and also over time in the same location, the average molecular composition is fairly uniform and is available from many sources, including NASA, which we use here.[1] We begin with the standard molar concentrations of dry air, then calculate the mass fractions by multiplying each mole fraction by the respective molecular weight and dividing by the sum. The results are shown in the following table:

Table 2. Composition of Dry Air

mol. wt.	by mole	by mass	gas
28.013	78.0843%	75.5188%	N_2
31.999	20.9476%	23.1416%	O_2
39.948	0.9367%	1.2919%	Ar
44.010	0.0314%	0.0477%	CO_2
28.965	100.0000%	100.0000%	total

Boilers operate on ambient air, which is not dry and contains some moisture (relative humidity >0). We next calculate the ambient humidity ratio, W, which is the mass ratio of water vapor to dry air:

---

[1] *CRC Handbook of Chemistry & Physics*, 1997 Edition (references NASA).

$$W = \frac{m_{water.vapor}}{m_{dry.air}} \qquad (2.10)$$

See Appendix E for more details on moist air properties. For the purposes of illustration, we will use an ambient temperature of 72°F, a relative humidity of 68%, and a barometric pressure of 14.54 psia. The humidity ratio for these conditions is 0.011565 (no units). Composition of moist air at these conditions is shown in the following table:

Table 3. Composition of Moist Air

mol. wt.	by mole	by mass	gas
28.013	76.6589%	74.6554%	N_2
31.999	20.5652%	22.8771%	O_2
39.948	0.9196%	1.2771%	Ar
44.010	0.0308%	0.0472%	CO_2
18.015	1.8254%	1.1432%	H_2O
28.7652	100.0000%	100.0000%	total

Complete Combustion

Stoichiometric is a chemical term generally referring to the ideal proportion of reactants. In the context of combustion, this term means just the right amount of oxygen (number of moles) to accommodate complete combustion (all oxidation states at their optimum level). For coal, fuel oil, or natural gas, this primarily means CO_2, H_2O, and SO_2. Ash is treated as a non-combustible. Hydroxl (OH), carbon monoxide (CO), sulfur monoxide (SO), and oxides of nitrogen (NOx) are undesirable emissions, corrosive, and mean failure to release all of the heat of combustion. Fire in a boiler can become unstable, even causing catastrophic explosions; therefore, boilers are always operated with *excess* air; that is, more than enough oxygen to provide complete combustion. Excess air may vary from 5% to 20%, depending on design and emission considerations. The air-to-fuel mass ratio is given the symbol, φ, and defined by:

$$\varphi = \frac{\dot{m}_{aur}}{\dot{m}_{fuel}} \qquad (2.11)$$

Balancing the Reaction

In chemistry class we often consider combustion of some fuel with oxygen alone, but this is rarely done in practice—perhaps for a space launch, but not a boiler. The coal-fired power plant just up from my house (Bull Run) burns more than 2½ railroad carloads of coal per hour or a whole trainload per day to heat 6,335,200 pounds of steam per hour (798 kg/s) at 3515 psia (24.2 MPa) to 1010°F (543°C). With an air/fuel mass ratio of approximately eight ($\varphi \approx 8$), this requires 7,721,833 lbm/hr (973 kg/s) of air! Clearly, burning pure oxygen is not an option. This means that we must consider the constituents of coal (Table 1) and also ambient air (Table 3). To balance the combustion equation, we consider the following reactions on a molar basis:

$$C + O_2 \rightarrow CO_2$$
$$H + \tfrac{1}{4}O_2 \rightarrow \tfrac{1}{2}H_2O \quad (2.12)$$
$$S + O_2 \rightarrow SO_2$$

Recall that the coal analysis (Table 1) was given in H (not H_2), in the solid (not gaseous) state. The complete balance must also consider the oxygen (as O, not O_2) included in the coal analysis plus the moisture in the coal (as H_2O) and ambient air (also as H_2O). The coefficients following the plus (+) in Equation 2.12 (i.e., 1, ¼, 1) are called *oxidation numbers*. We subtract one-half (-½) to account for the O (not O_2) in included in the coal analysis. These were listed in the first column of Table 1. We multiply these by the mole fractions and take the sum to arrive at the stoichiometric moles of O_2 required per mole of coal.

$$\omega = \sum_i n_i m_i \quad (2.13)$$

This can easily be accomplished using the SUMPRODUCT function in an Excel™ spreadsheet. Dividing this by the mole fraction of O_2 in the ambient air yields a stoichiometric *molar* air/fuel ratio of 2.24045 (no units). Multiplying this by the ratio of the air (28.7652 bottom left of Table 3) and coal (10.186 bottom left of Table 1) molecular weights yields the stoichiometric *mass* air/fuel ratio of 6.3268.

Flue Gas Composition

We can now calculate what the composition of the flue gas would be operating with the stoichiometric air/fuel ratio, assuming nothing leaks in or out of the flow streams. Using the symbol z for flue gas mass fractions, we have:

$$z_{N_2} N_2 + z_{O_2} O_2 + z_{CO_2} CO_2 + z_{H_2O} H_2O + z_{SO_2} SO_2 + z_{Ar} Ar \quad (2.14)$$

The exiting number of moles (and mass) of each element must match the entering number so that we have a molar (and mass) conservation equation for each element. For argon atoms, we have:

$$\frac{\dot{m}_{air} y_{Ar}}{MW_{Ar}} = \frac{\dot{m}_{flue} z_{Ar}}{MW_{Ar}} \quad (2.15)$$

Conservation of sulfur atoms:

$$\frac{\dot{m}_{coal} x_S}{MW_S} = \frac{\dot{m}_{flue} z_{SO_2}}{MW_{SO_2}} \quad (2.16)$$

Conservation of nitrogen atoms:

$$\frac{\dot{m}_{air} 2 y_{N_2}}{MW_{N_2}} + \frac{\dot{m}_{coal} x_N}{MW_N} = \frac{\dot{m}_{flue} 2 z_{N_2}}{MW_{N_2}} \quad (2.17)$$

Conservation of hydrogen atoms:

$$\frac{\dot{m}_{air} 2 y_{H_2O}}{MW_{H_2O}} + \frac{\dot{m}_{coal} x_H}{MW_H} = \frac{\dot{m}_{flue} 2 z_{H_2O}}{MW_{H_2O}} \quad (2.18)$$

Conservation of carbon atoms:

$$\frac{\dot{m}_{air} y_{CO_2}}{MW_{CO_2}} + \frac{\dot{m}_{coal} x_C}{MW_C} = \frac{\dot{m}_{flue} z_{CO_2}}{MW_{CO_2}} \quad (2.19)$$

Conservation of oxygen atoms (assuming complete combustion):

$$\dot{m}_{air}\left(\frac{2 y_{O_2}}{MW_{O_2}} + \frac{2 y_{CO_2}}{MW_{CO_2}} + \frac{y_{H_2O}}{MW_{H_2O}}\right) = $$
$$\dot{m}_{flue}\left(\frac{2 z_{O_2}}{MW_{O_2}} + \frac{2 z_{CO_2}}{MW_{CO_2}} + \frac{2 z_{SO_2}}{MW_{SO_2}} + \frac{y_{H_2O}}{MW_{H_2O}}\right) \quad (2.20)$$

In addition to these 6 equations (2.15-2.20), we also have the sum of mass fractions:

$$z_{N_2} + z_{O_2} + z_{CO_2} + z_{H_2O} + z_{SO_2} + z_{Ar} = 1 \quad (2.21)$$

and also the mass flow rate:

$$\dot{m}_{flue} = \dot{m}_{coal} - \dot{m}_{ash} + \dot{m}_{air} \quad (2.22)$$

These equations can be solved for the flue mass fractions. The equations for argon and SO_2 are trivial. The flue gas nitrogen mass fraction is:

$$z_{N_2} = \frac{\dot{m}_{air} y_{N_2} + \dot{m}_{coal} x_N}{\dot{m}_{flue}} \quad (2.23)$$

The flue gas CO2 mass fraction is:

$$z_{CO_2} = \frac{\left(\dfrac{\dot{m}_{air} y_{CO_2}}{MW_{CO_2}}\right) + \left(\dfrac{\dot{m}_{coal} x_C}{MW_C}\right)}{\left(\dfrac{\dot{m}_{flue}}{MW_{CO_2}}\right)} \quad (2.24)$$

The flue gas water vapor (H_2O) mass fraction is:

$$z_{H_2O} = \frac{\left(\dfrac{2\dot{m}_{air} y_{H_2O}}{MW_{H_2O}} + \dfrac{\dot{m}_{coal} x_H}{MW_H}\right)}{\left(\dfrac{2\dot{m}_{flue}}{MW_{H_2O}}\right)} \qquad (2.25)$$

The flue gas oxygen mass fraction is:

$$\frac{z_{O_2}}{MW_{O_2}} = \left\{ \frac{\dot{m}_{air}\left(\dfrac{y_{O_2}}{MW_{O_2}} + \dfrac{y_{CO_2}}{MW_{CO_2}} + \dfrac{y_{H_2O}}{2MW_{H_2O}}\right)}{\dot{m}_{flue}} - \left(\dfrac{z_{CO_2}}{MW_{CO_2}}\right) - \left(\dfrac{z_{SO_2}}{MW_{SO_2}}\right) - \left(\dfrac{z_{H_2O}}{2MW_{H_2O}}\right) \right\} \qquad (2.24)$$

Each of these equations can be modified to handle incomplete combustion, including the formation of carbon monoxide (CO), which often occurs with operating boilers and will be discussed subsequently. With these relationships, we can now calculate the stoichiometric flue gas composition for this particular coal with complete combustion using ambient air at the previously stated conditions.

Table 4. Flue Gas (Stoichiometric)

mol. wt.	by mole	by mass	gas
28.013	67.865%	65.043%	N_2
31.999	0.000%	0.000%	O_2
44.010	16.261%	24.484%	CO_2
18.015	15.010%	9.252%	H_2O
39.948	0.813%	1.111%	Ar
64.059	0.050%	0.110%	SO_2
29.229	100.000%	100.000%	total

The molecular weight of this mixture of gases is 29.229 (Table 4 bottom left) and is again calculated with the SUMPRODUCT function. The "by mass" column contains the z components from the preceding equations and the "by mole" column contains these divided by the respective molecular weights and also the sum.

Excess Air

If this boiler were operated with 18% excess air, then we can adjust the air/fuel ratio upward accordingly and recalculate the composition of the flue gas in the same manner. All of these calculations can be found in boiler1.xls in the examples folder.

Table 5. Flue Gas (18% excess air)

mol. wt.	by mole	by mass	gas
28.013	69.295%	66.584%	N_2
31.999	3.342%	3.669%	O_2
44.010	13.623%	20.566%	CO_2
18.015	12.867%	7.951%	H_2O
39.948	0.830%	1.138%	Ar
64.059	0.042%	0.092%	SO_2
29.154	100.000%	100.000%	total

The molecular weight of this flue gas is 29.154 (Table 4 bottom left). The combustion chemistry must always be calculated on a molar basis (i.e., mole-balanced); however, the specific heat and enthalpy must be calculated on a mass basis, as we will see in the next chapter, which is why we have calculated it both ways here.

Chapter 3. Heat Transfer

In order to calculate the heat transfer throughout the boiler, we must consider specific components and assign state points. These details vary from one design to another. The following figure is typical and will be used for these first sample calculations.

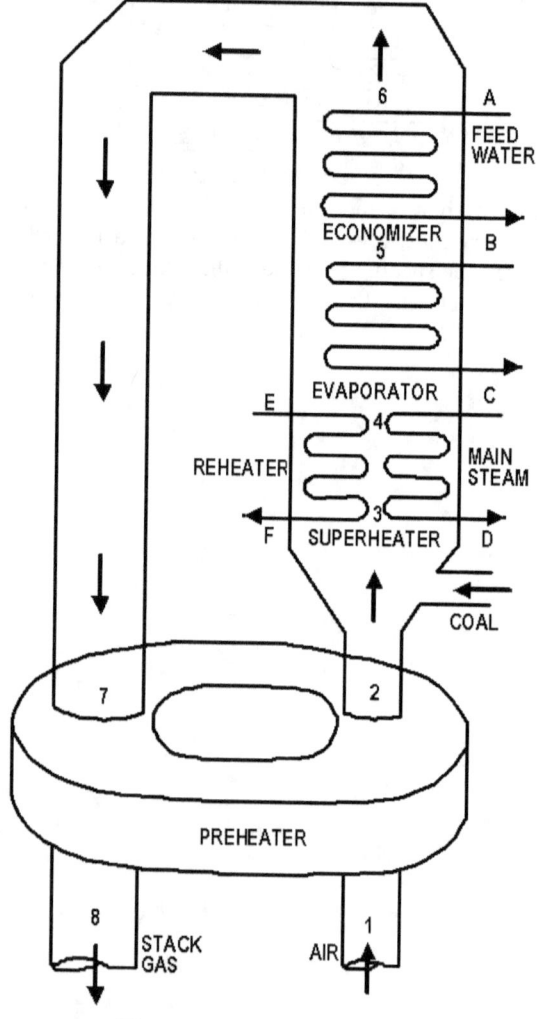

Figure 5. Boiler Schematic

Notice the counterflow arrangement of elements in Figure 5. This is optimal for heat transfer. While the relative orientation of each element may not be strictly counterflow, this is the ideal to achieve where practical.

Air Preheater

There are many designs for air preheaters. These must not only transfer heat, they must also handle the flue gas, which often contains soot and ash. One way of handling the soot problem is to use a rotating drum that heats as it rotates through the flue gas and cools as it rotates through the incoming air stream, effectively transferring the energy and preheating the air.

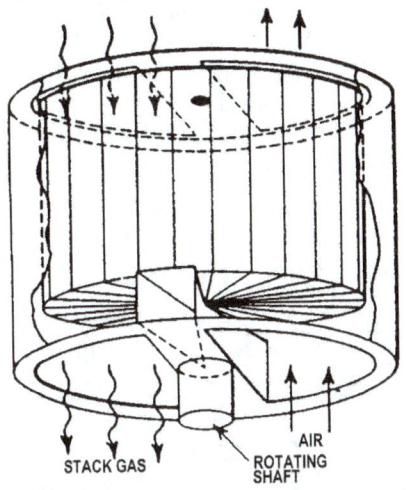

Figure 6. Air Preheater Schematic

Figure 7. Typical Air Preheater Drum

In Figure 5 the air path begins at 1 and proceeds through 8. There are 4 steam paths (economizer, evaporator, superheater, and reheater) with inlets and exits designated A through F. This is a simplified schematic; for example, the coal is pulverized and blown into the combustion chamber so that there is some additional air entering and participating in the process. There is also some leakage (in and out) for the air and flue gas in the preheater. We will consider all of these corrections in subsequent discussions.

For simplification, we will consider only two types of gas (ambient air before combustion and flue gas after combustion). We have the mass fractions for each type (Table 3 and Table 5, respectively). We use specific heats consistent with NASA Glenn in both cases (see appendix D). The VBA code for these properties can be found in boiler1.xls. The freebie steam property functions (available at the link beneath the Preface) are used for steam properties (see Appendix F for more details).

Steam Side Calculations

Some typical design parameters are used to create the current example (boiler1.xls). On the steam side, these include:

Table 6. Boiler Design Parameters

Value	Parameter
4,195,139	main steam flow rate [lbm/hr]
4393.7	feedwater inlet pres. [psia]
569.8	feedwater inlet temp. [°F]
3954.4	economizer pres. [psia]
880.4	economizer exit temp. [°F]
3515.0	main steam exit pres. [psia]
1110.0	main steam exit temp. [°F]
3,509,291	reheat flow rate [lbm/hr]
816.8	reheat inlet pres. [psia]
690.6	reheat inlet temp. [°F]
751.5	reheat exit pres. [psia]
1125.0	reheat exit temp. [°F]
671,335	coal feed rate [lbm/hr]
5,814,136	flue gas flow rate [lbm/hr]
660.0	air preheat temp. [°F]
300.0	preheater exit temp. [°F]

Using these parameters we can calculate the pressure drop and heat transfer for each section of the boiler. These are often part of the manufacturer's guarantees and are important details to accurately quantify when performing a test. Most modern power plants have permanent station instrumentation that continuously reports these vital temperatures, pressures, and flows.

It is fairly easy to verify the accuracy of temperature and pressure measurements, as additional connections are often provided so that temporary calibrated instruments can be attached at the same location. Measuring flow is more complicated and beyond the scope of this text; however, we will note that a precision calibrated orifice or nozzle (which can be inspected) is preferable. Clamp-on flow meters are so inaccurate as to be a complete waste of time and effort. Annubars are pretty useless too (don't believe the brochure). You might as well consult a medium or use a forked willow branch.

The enthalpies and sectional heat transfer corresponding to these design parameters are:

Table 7. Steam Side Calculations

97	steam property formulation
10.0%	economizer pressure drop
11.1%	superheater pressure drop
20.0%	combined pressure drop
8.0%	reheater pressure drop
568.5	main steam inlet enthalpy [BTU/lbm]
1292.0	economizer exit enthalpy [BTU/lbm]
1503.6	enthalpy at superheater exit [BTU/lbm]
1331.3	cold reheat enthalpy [BTU/lbm]
1583.1	hot reheat enthalpy [BTU/lbm]
3.035E+09	economizer heat transfer [BTU/hr]
8.879E+08	superheater heat transfer [BTU/hr]
8.835E+08	reheater heat transfer [BTU/hr]
4.807E+09	total heat transfer to steam [BTU/hr]

Note the "97" at the top left of Table 7. This spreadsheet (boiler1.xls) is set up so that you can pick which steam property formulation to use if you have the freebie Add-In. The formulation is often specified in the Contract. You will have enough details to harmonize without adding different steam properties to the mix. This spreadsheet has two tabs: one with English and the other with SI units. To switch steam properties from English to SI, simply replace 97 with -97.

In order to calculate the heat added by the coal, we use the composition (Table 1) along with the formulas described in Appendix B. For this coal we get the following using Equations B.1 and B.2:

Table 8. Calculated Heating Values

ht. val.	BTU/lbm	kJ/kg
HHV	8300	19,305
LHV	7680	17,864

The coal and ash have some sensible heat. Though this is small, we can correct for the associated heat addition from the coal and heat removal with the ash using a simple formula:

$$h = C_P \Delta T \qquad (3.1)$$

The constant pressure specific heat for water in English units is, of course, 1 BTU/lbm/°F, as this is the definition of a British Thermal Unit. The corresponding constant pressure specific heat of coal is approximately 0.3, making the combined result:

$$C_{Pcoal} = 0.3(1 - x_{H_2O}) + 1.0 x_{H_2O} \tag{3.2}$$

The constant pressure specific heat of ash is approximately 0.25 BTU/lbm/°F (for SI units multiply by 4.1868). Given the estimated respective temperatures, we get:

Table 9. Calculated Heats

	specific heat	
	BTU/lbm/°F	kJ/kg/°K
coal	0.52	2.16
ash	0.25	1.05
	sensible heat	
	BTU/lbm	kJ/kg
coal	6.2	14.4
ash	180.9	420.8

We can now calculate the total heat added by the coal and removed by the ash. The efficiency is the net heat transferred to the steam divided by that provided by the coal on an LHV basis, of course, because the H_2O in the flue is in the vapor (not liquid) state.

Table 10. Calculated Efficiency and Losses

5.156E+09	LHV heat input from coal [BTU/hr]
93.23%	boiler LHV efficiency
86.26%	boiler HHV efficiency
7.538E+08	heat transferred by preheater [BTU/hr]
3.395E+08	heat loss to flue [BTU/hr]
7.226E+06	boiler heat loss [BTU/hr]
7.226E+06	boiler heat loss check [BTU/hr]

The LHV efficiency is applicable here because H_2O in the flue is in the vapor state. Were the H_2O following combustion condensed to the liquid state, then the HHV efficiency would be applicable. The ratio of the two efficiencies is simply equal to the ratio of the heating values. The last two rows in Table 10 are often calculated as a check. The first is the sum of all the heat input streams minus the sum of all the heat output streams. The second is energy change in the flue pathway at the point of loss (lumped).

We next consider the temperatures and enthalpies along the flue pathway from the air inlet, through the preheater, the combustion, the steam tubes, back through the preheater, and out the stack, as shown in Figure 5. These values are listed in the following table:

Table 11. Temperatures and Enthalpies

Point	Temp. [°F]	Enthalpy [BTU/lbm]	
1	72.0	2.8	inlet air
2	660.0	148.4	preheated air
3	3356.1	1013.2	after combustion
4	2467.0	710.5	after superheater
6	783.7	191.8	after economizer
6A	784.0	191.8	after ash removal
7	779.5	190.6	before preheater
8	300.0	60.9	after preheater
A	569.8	568.5	feedwater
C	880.4	1292.0	after economizer
D	1110.0	1503.6	main steam
E	690.6	1331.3	cold reheat steam
F	1125.0	1583.1	hot reheat steam

The log-mean temperature difference and conductance ($Q=UA\Delta T$) along the path are listed in this next table

Table 12. Temperature Difference and Conductance

Points	LMTD	UA	
72/81	168.0	4.488E+06	preheater
3D/4C	1897.3	4.680E+05	superheater
3F/4E	1995.2	4.428E+05	reheater
4C/6A	685.1	4.430E+06	economizer
		9.829E+06	total for boiler
60/70	709.7	1.018E+04	heat loss

We have calculated an effective heat loss (last item in Table 12) and corresponding temperature difference, ΔT, and conductance, UA. When we consider losses (which are an essential part of PTC4), this will be useful for comparison.

Chapter 4. Uncertainty

The primary document on boiler performance testing is ASME PTC4[2], which I have followed for decades as it has evolved. The reaction of newcomers to this approach is often, "What are all these loss calculations for? Why not just measure everything and calculate the results directly?" Ah, yes... that is the question. We will discuss instrument accuracy later, but that's not the real problem.

Even if you could accurately measure the coal flow up a conveyor belt and even if none of it fell off before being pulverized and blown into the boiler, you'd still be left with the fact that the coal is not uniform and no practical amount of sampling will resolve this uncertainty. "Then just go to the other end," is often the response when the problem of coal measurement is explained. Flue (outlet gas) measurement is also problematic. Not only does the gas velocity, temperature, and molar composition change spatially over the flow area, but these also vary over time—and not so slowly that temporal averaging is adequate. Representative temperature is necessary. Not only is this required to calculate enthalpy, we must also calculate density to go from velocity measurements to mass flow rate.

Accurately measuring any one of these items throughout a test, let alone on a continuous basis, is a daunting task. The most precise instruments are also the most susceptible to fouling and corrosion. Maintaining an array of temperature sensors across a boiler stack would be a full-time job. An array of velocity sensors wouldn't last more than a few days and would quickly drift away from calibration. Such measurements are most often derived from a single instrument, which is durable (though not as accurate as a delicate one) and strategically positioned to hopefully be representative of the whole (though inherently less accurate than an array of instruments).

While considerable effort is often made to measure each of these things (coal flow, coal composition, exhaust flow, exhaust temperature, and exhaust composition), a different approach can yield acceptable results with far less effort. The reasoning is: If we can quantify a term (for example radiative heat loss) to within ±10% that is itself only 10% of the total sought, then the impact on the whole is only ±1%. PTC4 breaks down the total process into smaller pieces and deals with these individually and empirically; based on many years of experience to obtain a more accurate estimate of the whole than could be achieved without a less practical measurement effort.

To illustrate this concept we will build a second spreadsheet (boiler2.xls) that is similar in many ways to the first (boiler1.xls) but focuses on the quantities that can be accurately measured and the ones that are estimated. This

---

[2] American Society of Mechanical Engineers, Performance Test Code 4, Fired Steam Generators.

will be similar (though not exactly like) the example in Appendix B of PTC4-2008. We begin with the ambient conditions in SI units and coal composition:

Table 13. Ambient Conditions

0.1000	barometric pres. [MPa]
298.4	ambient dry-bulb temp. [°K]
58%	ambient relative humidity

Table 14. Coal Composition

ox. num.	mol. wt.	by mass	by mole	
1.00	12.011	55.02%	35.630%	C
0.25	1.008	3.61%	4.860%	H
-0.50	15.999	7.81%	0.629%	O
0.00	14.007	0.89%	0.365%	N
1.00	32.060	1.18%	2.427%	S
0.00	51.746	12.61%	10.452%	Ash
0.00	18.015	18.90%	35.630%	H_2O
0.52479	9.962	100.00%	100.00%	total

In the previous example (Chapters 1 and 2) we began with the coal and flue mass flow rates. In this chapter we explained how these would be very difficult to measure accurately. We can fairly easily and accurately measure the excess O_2 at the stack. Most modern coal plants have a Continuous Emission Monitoring System (CEMS) that reports stack O_2. With considerable turbulence and available fuel, this is somewhat uniformly distributed across the stack so that a single measurement is often adequate.

This is fortuitous, because we can use the relationship along with a calculated radiative heat loss plus the accurately measured steam flow to calculate the coal and combustion air flow rates based on the coal composition. If we can measure the coal composition and if it is fairly constant and if we can estimate the radiative heat loss and if we can measure the excess O_2 at the stack, then we have enough equations and unknowns to solve the combustion problem. This method of calculation has been championed by Professor Sastry Munukutla and others, as described in many publications.[3,4,5]

---

[3] Levy, E.K., S. Munukutla, A. Jibilian, H.G. Crim, J.G. Cogoli, A.F. Kwasnik, and F. Wong, "Analysis of the Effects of Coal Fineness, Excess Air and Exit Gas Temperature on the Heat Rate of a Coal Fired Power Plant, 84-JPGC-PWR-1, ASME Joint Power Conference, 1984.

[4] Levy, E. S. Munukutla, O. Badr, S. Williams, and J. Fernandes, "Optimization of Unit Heat Rate Through Variations in Fireside Parameters," Power Plant Performance Monitoring and System Dispatch Conference, 1986.

[5] Levy, E., N. Sarnac, H.G. Crim, R. Leyse, and J. Lamont, "Output/Loss: A New Method for Measuring Unit Heat Rate" 87-JPGC-PWR-39, ASME Joint Power Generation Conference, 1987.

While calculating the radiation heat loss might seem like a daunting task, boiler manufacturers often provide considerable information to assist in this process, including: pertinent sections, sectional areas, and typical temperatures. Formulas are provided in PTC4 to estimate applicable heat transfer coefficients. I have even used infrared photography to infer skin temperatures and applied these to a finite element representation of the boiler, such as the following:

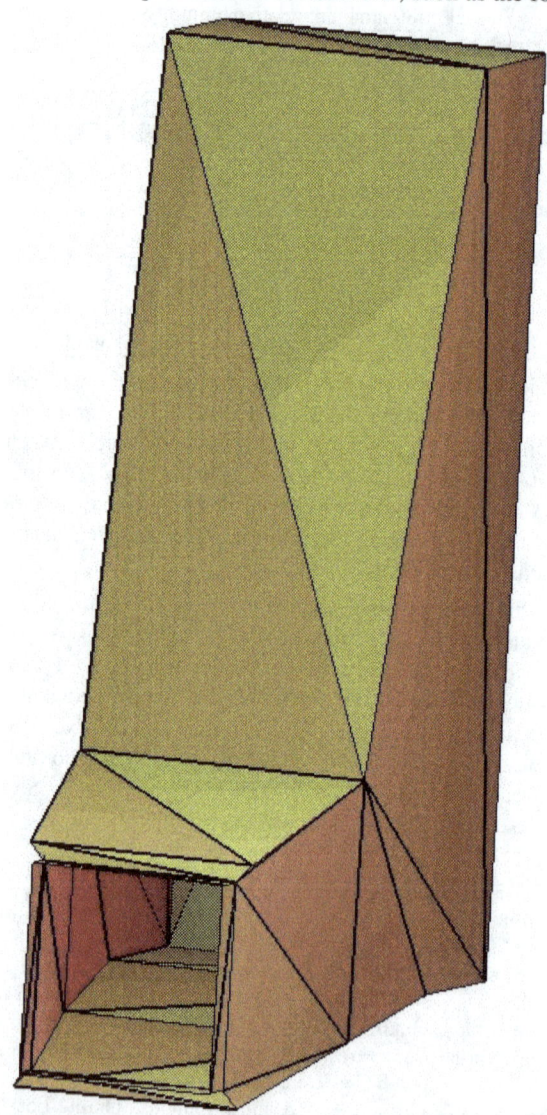

Figure 8. Boiler Surface Broken Down Into Elements

For this particular boiler, the manufacturer provided an effective surface area and heat transfer coefficient: $UA = 22.42$ kW/°K (42,500 BTU/hr/°F or 80,712 kJ/hr/°K).

While calculating the mass fraction of oxygen in the flue gas (Equation 2.24) we utilized the oxidation numbers (column 1 of Table 1). Oxidation numbers are the number of moles of O_2 required to completely react with the particular constituent. Oxidation numbers will be quite useful when considering combustion of natural gas, which has many more constituents. From these we get the stoichiometric air/fuel ratio, φ_s.

To simplify the equations, we define several intermediate variables:

$$\dot{Q}_{steam} = \dot{m}_{steam}(h_{ms} - h_{fw}) + \dot{m}_{reheat}(h_{hrh} - h_{crh}) \quad (4.1)$$

where h_{ms} and h_{fw} are the enthalpies of the main steam at the exit of the superheater and inlet from feedwater, respectively, and h_{hrh} and h_{crh} are the enthalpies of the hot and cold reheat steam, respectively.

$$\dot{Q}_{rad} = UA_{rad}(T_{rad} - T_{db}) \quad (4.2)$$

where UA_{rad} is the effective overall conductance (which may be on a sectional basis). T_{rad} and T_{db} are the effective radiative temperature (which may be on a sectional basis) and the ambient dry-bulb temperature.

$$\Delta h_{stk} = (h_{stack} - h_{amb})(1 + \varepsilon)\varphi_s \quad (4.3)$$

where h_{stack} is the enthalpy of the flue gas exiting the boundary and h_{amb} is the enthalpy of the ambient moist air, ε is the excess air, and φ_s is the stoichiometric mass air/fuel ratio (which we calculated from the coal composition and oxidation numbers).

$$\Delta h_{ash} = (h_{stack} - h_{ash})x_{ash} \quad (4.4)$$

where h_{ash} is the sensible heat of the ash, and x_{ash} is the mass fraction of ash in the coal.

$$HV = LHV + h_{coal} - h_{stack} \quad (4.5)$$

where HV is the adjusted heating value and h_{coal} is the sensible heat of the coal. Note that we must use the LHV here because the NASA Glenn gas properties are based on water in the vapor state. We can now calculate the mass flow rate of coal explicitly:

$$\dot{m}_{coal} = \frac{\dot{Q}_{steam} + \dot{Q}_{rad}}{HV + \Delta h_{ash} + \Delta_{hstk}} \quad (4.6)$$

Once we have the coal mass flow rate, we can easily calculate the combustion air and flue mass flow rates. Steam flow, pressures, and temperatures for this boiler are listed in the following table:

Table 15. Steam Parameters

Value	Parameter
2,563,242	main steam flow rate [kg/hr]
31.34	feedwater inlet pres. [MPa]
556.5	feedwater inlet temp. [°K]
26.54	main steam exit pres. [MPa]
841.5	main steam exit temp. [°K]
2,130,143	reheat flow rate [kg/hr]
4.52	cold reheat pres. [MPa]
584.3	cold reheat temp. [°K]
4.33	hot reheat pres. [MPa]
852.6	hot reheat temp. [°K]

Additional design parameters include the following, some of which would be measured during a test:

Table 16. Additional Design/Test Parameters

Value	Parameter
304.8	coal temp. [°K]
428.2	ash temp. [°K]
448.2	stack temp. [°K]
692.6	average surface temp. [°K]
80,712	radiation loss UA [kJ/hr/°K]
17%	excess Air

As before, we have the parameters calculated from the coal analysis and mass fractions of the moist ambient air:

Table 17. Computed Parameters

Value	Parameter
2.55291	stoichiometric molar air/fuel ratio
7.37027	stoichiometric mass air/fuel ratio
8.62322	operating A/F ratio
577,278	stoichiometric O2 flow rate [kg/hr]
2,524,059	stoichiometric moist air flow rate [kg/hr]
22,839	HHV [kJ/kg]
21,615	LHV [kJ/kg]
1.81	Cp coal [kJ/kg°K]
1.05	Cp ash [kJ/kg°K]
29.2	SH coal [kJ/kg]
146.0	SH ash [kJ/kg]

The steam calculations are shown in this next table (notice the minus 97 indicating SI properties):

Table 18. Steam Parameters

-97	steam property formulation
1245.6	main steam inlet enthalpy [kJ/kg]
3380.9	main steam exit enthalpy [kJ/kg]
2975.8	cold reheat enthalpy [kJ/kg]
3624.8	hot reheat enthalpy [kJ/kg]
5.473E+09	main steam heat transfer [kJ/hr]
1.383E+09	reheater heat transfer [kJ/hr]
6.856E+09	total heat transfer to steam [kJ/hr]

After calculating the ambient and stack enthalpies using NASA Glenn, we next calculate the coal mass flow rate using Equation 4.6, then the ash, combustion air, and stack mass flow rates. The LHV efficiency is the rate of heat transferred to the steam divided by the total. The HHV efficiency is equal to the LHV efficiency times the ratio of the heating values.

Table 19. Calculated Flows and Efficiencies

9.6	ambient air enthalpy [kJ/kg]
168.1	stack enthalpy [kJ/kg]
342,465	coal feed rate [kg/hr]
43,169	ash flow rate [kg/hr]
2,953,149	inlet air flow rate [kg/hr]
3,252,445	flue gas flow rate [kg/hr]
7.402E+09	LHV heat input from coal [kJ/hr]
92.62%	boiler LHV efficiency
87.66%	boiler HHV efficiency

As a check, we calculate the mass and energy flow rates for each process (which we have lumped together as indicated in the tables) entering and leaving then take the difference to make sure the mass and energy match.

Table 20. Mass and Heat Inputs

kg/hr	kJ/hr	fraction	source
2.953E+06	2.845E+07	0.168%	from ambient air
3.425E+05	7.412E+09	43.673%	from coal (combustion + sensible)
4.693E+06	9.532E+09	56.159%	from steam
7.989E+06	1.697E+10	100.000%	total

The corresponding mass and energy outputs are:

Table 21. Mass and Heat Outputs

kg/hr	kJ/hr	fraction	destination
4.693E+06	1.639E+10	96.554%	to steam
3.252E+06	5.467E+08	3.221%	through stack
4.317E+04	6.301E+06	0.037%	with ash
0.000E+00	3.181E+07	0.187%	radiation loss
7.989E+06	1.697E+10	100.000%	total
0.000E+00	0.000E+00		error

Considering the complexity of these calculations (which may include unit conversions) it is good practice to include several checks in the spreadsheet.

Chapter 5. Gas and Oil Firing

To illustrate handling fuels other than coal, we will modify the spreadsheet from the previous chapter to form a new similar one (boiler3.xls), also with both English and SI units. Because natural gas often contains some helium, we will also expand our composition of ambient air. Another difference with this third spreadsheet is that it contains the 1967 steam formulation implemented in VBA (Visual Basic for Applications, the Microsoft™ script language for Excel™).

The ambient conditions for this boiler are:

Table 22. Ambient Conditions

14.540	amb. pres.[psia]
95	amb. temp.[°F]
43%	amb. rel. hum.

The dry air composition (with helium) is listed in this next table:

Table 23. Dry Air Composition

MW	by mole	by mass	gas
28.0134	78.08422%	75.5192%	N_2
31.9988	20.94766%	23.1418%	O_2
39.9480	0.93594%	1.2908%	Ar
44.0095	0.03160%	0.0480%	CO_2
4.0026	0.00057%	0.0001%	He
28.9649	100.00000%	100.0000%	total

At these conditions the ambient humidity ratio, W, is 0.01544 from which we calculate the moist air composition:

Table 24. Moist Air Composition

MW	by mole	by mass	gas
28.0134	76.19299%	74.3711%	N_2
31.9988	20.44030%	22.7900%	O_2
39.9480	0.91327%	1.2712%	Ar
44.0095	0.03083%	0.0473%	CO_2
4.0026	0.00056%	0.0001%	He
18.0153	2.42204%	1.5204%	H_2O
28.6997	100.0000%	100.0000%	total

Note the molecular weight of dry air at the bottom left of Table 23 and for moist air at the bottom left of Table 24.

While natural gas may have over 40 constituents, we will only consider 17 common ones in this example. Unlike coal composition, which is specified in mass fractions, natural gas composition is specified in mole (or volume) fractions. Natural gas composition is usually measured by a gas chromatograph. The composition for this example is listed in the following table:

Table 25. Natural Gas Composition

MW	by mole	by mass	gas
16.0425	79.58%	63.5837%	Methane
30.0690	9.92%	14.8616%	Ethane
44.0956	1.75%	3.8476%	Propane
58.1222	0.38%	1.0884%	IsoButane
58.1222	0.37%	1.0768%	Butane
72.1488	0.15%	0.5462%	IsoPentane
72.1488	0.10%	0.3450%	Pentane
86.1754	0.25%	1.0644%	Hexane
28.0134	2.04%	2.8517%	Nitrogen
28.0101	0.28%	0.3906%	Carbon Monoxide
44.0095	4.28%	9.3898%	Carbon Dioxide
18.0153	0.09%	0.0808%	Water
34.0809	0.44%	0.7468%	Hydrogen Sulfide
2.0159	0.27%	0.0271%	Hydrogen
4.0026	0.04%	0.0080%	Helium
31.9988	0.02%	0.0319%	Oxygen
39.9480	0.03%	0.0597%	Argon
17.0311	100.00%	100.0000%	total

Note the molecular weight in the lower left corner of the table calculated using the SUMPRODUCT function.

Several important quantities can be readily calculated using the SUM and SUMPRODUCT functions in an Excel spreadsheet, provided the parameters are correctly arranged in tabular form. We have already seen this with molecular weights and even stoichiometric O_2 ratio. With this more complex fuel, we will also use these spreadsheet functions to sum up the number of moles (and mass) of each of the elements (carbon, hydrogen, oxygen, sulfur, nitrogen, argon, and helium). The following tabular form is quite useful:

Table 26. Gas Properties

name	formula	BTU/lbm		mole factors							
		LHV	HHV	C	H	N	O	S	Ar	He	O_2
Methane	CH_4	21,511	23,891	1	4	0	0	0	0	0	2.0
Ethane	C_2H_6	20,428	22,332	2	6	0	0	0	0	0	3.5
Propane	C_3H_8	19,921	21,653	3	8	0	0	0	0	0	5.0
IsoButane	C_4H_{10}	19,590	21,232	4	10	0	0	0	0	0	6.5
Butane	C_4H_{10}	19,657	21,299	4	10	0	0	0	0	0	6.5
IsoPentane	C_5H_{12}	19,455	21,043	5	12	0	0	0	0	0	8.0
Pentane	C_5H_{12}	19,497	21,085	5	12	0	0	0	0	0	8.0
Hexane	C_6H_{14}	16,983	18,519	6	14	0	0	0	0	0	9.5
Nitrogen	N_2	0	0	0	0	2	0	0	0	0	0.0
Carbon Monoxide	CO	4,342	4,342	1	0	0	1	0	0	0	0.5
Carbon Dioxide	CO_2	0	0	1	0	0	2	0	0	0	0.0
Water	H_2O	0	0	0	2	0	1	0	0	0	0.0
Hydrogen Sulfide	H_2S	6,534	7,094	0	2	0	0	1	0	0	1.5
Hydrogen	H_2	51,539	61,026	0	2	0	0	0	0	0	0.5
Helium	He	0	0	0	0	0	0	0	0	1	0.0
Oxygen	O_2	0	0	0	0	0	2	0	0	0	-1.0
Argon	Ar	0	0	0	0	0	0	0	1	0	0.0
per mole				1	2	2	1	1	1	1	1

Using the SUMPRODUCT function combining the third column in Table 25 with the third column in Table 26, we get an LHV of 18,339 BTU/lbm and an HHV of 20,274 BTU/lbm. The number of atoms per mole is on the bottom row (C, H_2, O_2, S, Ar, He), which we will also use to facilitate adding up constituent formulas.

Next we take the mole fractions of the fuel gas, multiply these by the bottom row in Table 26, and populate a column of C, H, N, O, S, Ar, and He. We guess some initial value of stoichiometric molar air/fuel ratio (8 to 15), combine this with the mole fractions of moist air, breaking each up into the number of elements, and add each to the corresponding cell with the fuel. This gives us the number of moles of each exhaust gas component (N_2, O_2, Ar, CO_2, He, H_2O, and SO_2).

For stoichiometric combustion, the moles of O_2 in the products are zero. We could just use the Excel Solver function to adjust the air/fuel ratio so as to obtain zero in this cell. If you haven't already used this, you must go to Add-Ins, Go..., and check the box next to Solver, as shown below:

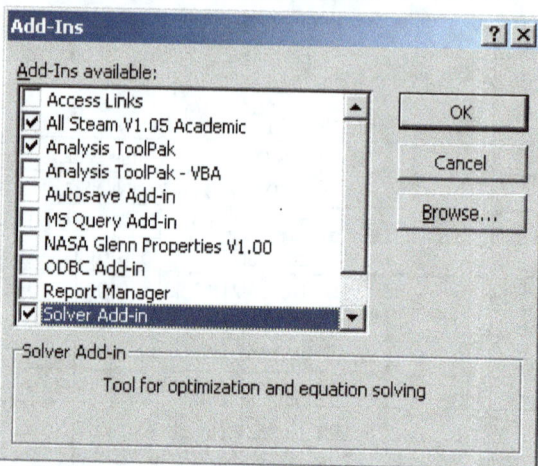

The solver dialog will look like this:

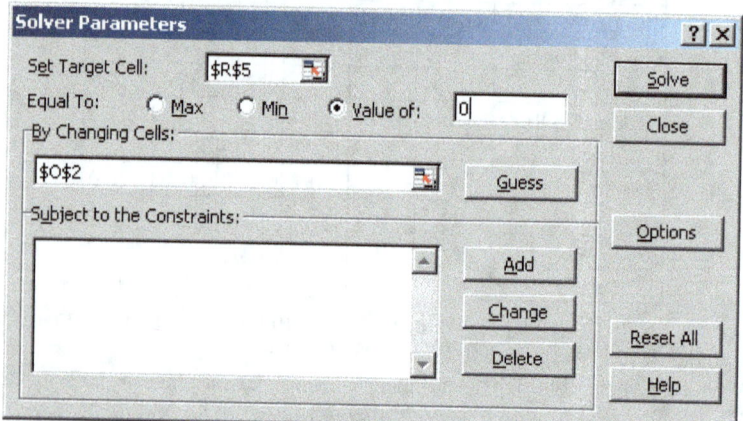

Or we could solve the problem algebraically for the stoichiometric molar air/fuel ratio:

$$\varphi_{\substack{stoic \\ molar}} = \frac{C_{fuel} + \frac{1}{4}H_{fuel} + S_{fuel} + Air_{SO_2} - \frac{1}{2}O_{fuel}}{Air_{O_2}} \quad (5.1)$$

where C_{fuel}, H_{fuel}, S_{fuel}, and O_{fuel} are the respective element counts we just calculated from the fuel composition and numbers in the bottom of Table 26. Air_{SO2} and Air_{O2} are the mole fractions of SO_2 and O_2 in the moist ambient air, respectively. To get the operating mix (with 15% excess air), we simply multiply by one plus the excess and calculate another column using the higher air/fuel ratio. The mole and mass fractions are then calculated as before. The results are shown in this next table:

Table 27. Molar Air/Fuel Ratio Calculations

moles		10.4096	11.9710	flue	per mole of fuel		flue gas composition		
elem	fuel	stoic	excess	gas	stoic.	excess	MW	by mole	by mass
C	1.1497	0.0032	0.0037	N_2	7.9723	9.1620	28.0134	70.067%	70.494%
H	4.0742	0.5042	0.5799	O_2	0.0000	0.3192	31.9988	2.441%	2.805%
N	0.0409	7.9314	9.1211	Ar	0.0954	0.1096	39.9480	0.838%	1.203%
O	0.0898	4.5140	5.1911	CO_2	1.1529	1.1534	44.0095	8.821%	13.942%
S	0.0044	0.0000	0.0000	He	0.0005	0.0005	4.0026	0.004%	0.001%
Ar	0.0003	0.0951	0.1093	H_2O	2.2892	2.3270	18.0153	17.796%	11.514%
He	0.0004	0.0001	0.0001	SO_2	0.0044	0.0044	34.0809	0.034%	0.041%
					11.5146	13.0760	27.8436	100.000%	100.000%

The four left columns are by element and the six right columns are by compound (gas). The stoichiometric molar air/fuel ratio is at the top left (10.4096) and the operating ratio (with excess air) is just to the right (11.9710).

There is only one steam path for this particular boiler (no reheat steam), as it is at a paper mill. The steam conditions are:

Table 28. Steam Parameters

375,500	main steam flow rate [lbm/hr]
1250	feedwater inlet pres. [psia]
405	feedwater inlet temp. [°F]
1100	steam exit pres. [psia]
1005	steam exit temp. [°F]

We will use the same approach as in Chapter 4, where we will estimate the losses and calculate all of the input and output energy streams to get the actual flow rates of fuel and air. There is no ash with this fuel but there will be some sensible heat associated with the fuel. To make this calculation, we have expanded the exhaust gas function and corresponding second-order regressions to accommodate the gas constituents (see boiler3.xls Alt-F11) so that we are using the same enthalpy calculations (and associated references) for the ambient moist air, fuel gas, and flue gas, assuring that there are no reference discrepancies.

> *Beware: Avoiding reference discrepancies when there is steam or water injection into the combustion stream (as is the case with some combustion gas turbines to control emissions) requires additional effort.*

The additional parameters needed to solve this boiler are listed in the following table:

Table 29. Additional Design Parameters

50	fuel supply temp. [°F]
350	stack temp. [°F]
600	average surface temp. [°F]
3,000	radiation loss UA [BTU/hr/°F]
15%	excess Air

Again, we could use the Solver function to find the fuel mass flow rate that makes the entering and exiting energy equal or we can solve for it algebraically.

$$\dot{m}_{fuel} = \frac{UA(T_{rad} - T_{amb}) + \dot{m}_{steam}(h_{SH} - h_{FW})}{\varphi i(h_{air} - h_{flue}) + LHV + h_{fuel} - h_{flue}} \tag{5.2}$$

where φ is the operating (with excess air) mass air/fuel ratio. This is equal to the operating molar air/fuel ratio times the ratio of the air and fuel molecular weights; UA is the radiative loss conductance; T_{rad} is the effective radiative loss temperature; T_{amb} is the ambient (dry-bulb) temperature; m_{steam} is the steam mass flow rate, h_{SH} is the enthalpy of the exiting superheated steam; h_{FW} is the enthalpy of the entering feedwater; h_{AIR} is the enthalpy of the ambient air (at the dry-bulb temperature and humidity); h_{FLUE} is the enthalpy of the exiting flue gas; LHV is the composite lower heating value of the fuel; and h_{FUEL} is the sensible heat (enthalpy) of the composite fuel gas. The details are shown in the table below:

Table 30. Mass and Energy Balance

Inputs			
lbm/hr	BTU/hr	fraction	source
4.244E+05	1.860E+06	0.310%	from ambient air
2.480E+04	4.545E+08	75.794%	from fuel (LHV+sensible)
3.755E+05	1.433E+08	23.896%	from steam
8.247E+05	5.997E+08	100.000%	total
Outputs			
lbm/hr	BTU/hr	fraction	destination
3.755E+05	5.652E+08	94.255%	to steam
4.492E+05	3.294E+07	5.492%	through stack
0.000E+00	1.515E+06	0.253%	radiation loss
8.247E+05	5.997E+08	100.000%	total
0.000E+00	0.000E+00		error

The heat transferred to the steam and from the fuel, along with the LHV and HHV efficiency are shown in this next table:

Table 31. Overall Results

4.219E+08	heat transfer to steam [BTU/hr]
4.549E+08	heat input from fuel [BTU/hr]
92.76%	boiler LHV efficiency
83.90%	boiler HHV efficiency

If the fuel oil analysis is provided on a carbon, hydrogen, etc. basis, then use the coal boiler spreadsheet (either boiler1.xls or boiler2.xls). If it is provided on a distillate basis (by compound), then use the gas spreadsheet (boiler3.xls) and replace the methane, ethane, butane, propane, etc. with the appropriate constituents and corresponding properties.

Chapter 6. Incomplete Combustion

The most common product of incomplete combustion in boilers (which normally operate at near-atmospheric pressure) is carbon monoxide (CO). Formation of oxides of nitrogen (NOx) and sulfur (SOx), which are a concern for emissions, require higher pressures and so are found in the exhaust of reciprocating internal combustion (car and truck gasoline and diesel) engines and combustion gas turbines, which may compress the air up to 40 atmospheres before entering the combustion chamber.

We already added the enthalpy of CO to the VBA function in the previous example (boiler3.xls) because it can be found in natural gas. We also have the heating values (in the same spreadsheet) so that we can calculate the loss of potential heating due to incomplete combustion of carbon in the fuel. Combustion of CO to CO_2 yields 4342 BTU/lbm or 10,100 kJ/kg. As carbon monoxide is toxic, boiler operators are required to monitor it so that we have this information when evaluating performance.

Continuous Emission Monitoring System (CEMS) also report O_2, CO_2, and SO_2. Returning to the problem raised in Chapter 4, we now consider the relative accuracy of these stack gas measurements. While it may be quite problematic to measure coal and stack mass flow rates and somewhat less problematic to measure coal composition, it is fairly easy to measure these gases in the stack. The question then becomes, "Which is most accurate?" A well designed and maintained CEMS system is much more accurate than coal scales and even stack gas velocity measurements. It is at least as accurate as grab sampling of coal and some would argue that these measurements are more accurate.

In the previous examples we have the same number of equations and unknowns (balancing elements, mass, and energy). If we add CEMS measurements, we now have more information than is strictly necessary from an algebraic point of view. How do we approach this? By finding the *best* solution. What constitutes the best solution? The one with the smallest residual. There are many ways of calculating residuals. We will use the sum of the squares of the respective errors:

$$r^2 = \sum (x_{calculated} - x_{measured})^2 \qquad (6.1)$$

The first unknown involving any significant algebra that we calculated in the previous example was the stoichiometric molar air/fuel ratio. With a CEMS we are measuring O_2 in the stack so we will inherently be considering excess air if we obtain a best match. Here we seek the operating mass air/fuel ratio that best matches the values of O_2, CO_2, CO, and SO_2 reported by the CEMS. Finding the best value of air/fuel ratio becomes a problem of nonlinear minimization in one variable. The natural choice is Brent's method (see Appendix G). This has been programmed in VBA to facilitate use in the spreadsheet (boiler4.xls).

The algorithm consists of several main steps (the outer loop), including: initialize moist air properties based on ambient conditions, adjust estimated coal composition and separate ash, initialize stack (flue) gas composition, estimate air/fuel ratio then iteratively adjust using Brent's method to achieve the lowest residual, and calculate subsequent parameters, returning these in an array.

Table 32. Outer Loop

step	action
1	begin
2	moist air
3	adjust coal
4	separate ash
5	initialize stack
6	estimate air/fuel ratio
7	refine air/fuel ratio
8	calculate parameters
9	return array

The algorithm also has an inner loop that is repeatedly called by the Brent's method subroutine. This inner subroutine uses the current estimate of air/fuel ratio to balance the chemical reaction and return the residuals (discrepancies between the calculated mole fractions and those reported by the CEMS). The inner loop also has several steps.

Table 33. Inner Loop

step	action
1	balance reaction
2	stack gas composition
3	calculate errors
4	return residual

The core subroutine is listed below. The complete code can be found in spreadsheet boiler4.xls (press Alt-F11).

```
Function AirFuelRatio(W As Double, c As Double, H As
    Double, S As Double, O As Double, N As Double,
    FM As Double, Ash As Double, O2 As Double,
    CO2 As Double, CO As Double, SO2 As Double)
    As Variant
    Dim v As Variant, A As Double, B As Double,
    x As Double, y As Double
    v = MoistAir(W)
    v = CoalNet(c, H, S, O, N, FM, Ash, 1)
    stack_dry.O2 = O2
    stack_dry.CO2 = CO2
    stack_dry.CO = CO
    stack_dry.SO2 = SO2
```

```
x = coal_gross.C + coal_gross.H + coal_gross.S
    + coal_gross.O + coal_gross.N
y = coal_gross.FM + coal_gross.Ash
A = Brent(12, 16)
B = Combustion(A)
AirFuelRatio = Array(A * x / (x + y), Sqr(B),
    stack_wet.N2, stack_wet.O2, stack_wet.CO2,
    stack_wet.CO, stack_wet.H2O, stack_wet.Ar,
    stack_wet.SO2)
End Function
```

As with previous examples, we begin with certain inputs, including: ambient conditions, measured steam data, and CEMS measurements:

Table 34. Inputs

\multicolumn{2}{c}{Ambient Conditions}	
14.68	barometric pres. [psia]
57.5	ambient dry-bulb temp. [°F]
71%	ambient relative humidity
\multicolumn{2}{c}{Steam Data}	
2,857,650	main steam flow rate [lbm/hr]
2709	feedwater inlet pres. [psia]
511	feedwater inlet temp. [°F]
2408	main steam exit pres. [psia]
1002	main steam exit temp. [°F]
2,536,525	reheat flow rate [lbm/hr]
531	cold reheat pres. [psia]
630	cold reheat temp. [°F]
499	hot reheat pres. [psia]
1000	hot reheat temp. [°F]
\multicolumn{2}{c}{CEMS Data (dry molar basis)}	
2.7087%	O_2
14.0640%	CO_2
0.0034%	CO
0.0897%	SO_2

We must also have an estimated coal composition, which we will refine later. As before, this is on a mass basis (wet):

Table 35. Initial Coal Composition (wet)

mol. wt.	by mass	by mole	
12.011	69.44%	48.887%	C
1.008	5.06%	42.447%	H
15.999	4.89%	2.583%	O
14.007	1.19%	0.718%	N
32.060	1.30%	0.342%	S
51.746	11.38%	1.860%	Ash
18.015	6.74%	3.164%	H_2O
8.455	100.00%	100.00%	total

The ambient humidity ration for these conditions is 0.00719. This plus the values in Tables 34 and 35 are inputs to the calculation algorithm, which produces the following outputs:

Table 36. Algorithm Outputs

quantity	by mole	by mass	mol. wt.
AFR	11.257	N/A	N/A
R^2	0.3414%	N/A	N/A
N_2	70.1839%	62.7754%	28.013
O_2	3.1295%	3.1974%	31.999
CO_2	21.1099%	29.6633%	44.010
CO	0.0032%	0.0029%	28.010
H_2O	4.1598%	2.3928%	18.015
Ar	1.1993%	1.5297%	39.948
SO_2	0.2144%	0.4386%	64.059
total	100.0000%	100.0000%	31.319

As the function returns 9 values in an array, we arrange the values in this same order to facilitate calculation and eliminate redundancies. The corresponding mass fractions for the stack (flue) gas are calculated from the molecular weights and mole fractions as before.

The residual (square root of the sum of the squares of the discrepancies between calculated O_2, CO_2, CO, SO_2 and those reported by the CEMS) for these inputs is 0.3414%. We can further adjust the coal analysis to exactly match the CEMS, while simultaneously updating the air/fuel ratio. As we have four inputs (O_2, CO_2, CO, and SO_2) we can adjust four of the coal inputs. We also require that the sum is equal to 1. Optimal results are obtained by adjusting the smaller components (H, O, N, and S) and calculating the carbon (typically the largest component) from the sum=1 requirement, while holding the ash and moisture constant or the ratio of these to the carbon constant. One of these must be S; otherwise, we cannot exactly match reported SO_2.

To accomplish this refinement of the coal composition, we use the Excel Solver, as discussed in Chapter 5. We select the cell containing R^2 and check (·)Min, indicating we want to minimize this value. Then we select the four cells containing the H, O, N, and S values for the refined coal composition. The function quickly obtains a zero residual and corresponding results.

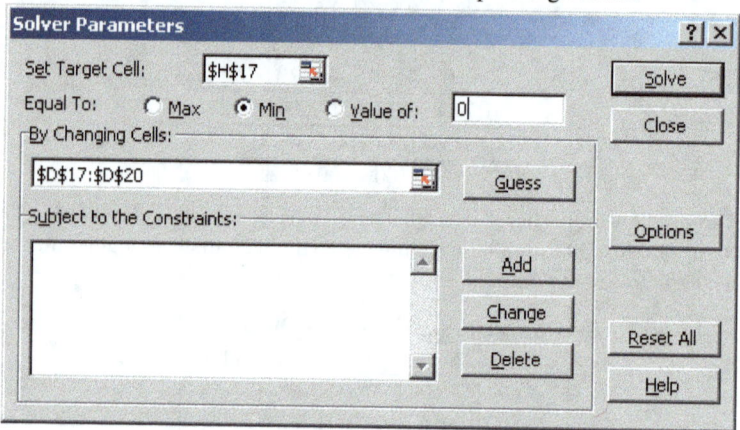

The refined (or adjusted) algorithm output is shown below (note R^2 is now zero):

Table 37. Refined Results

quantity	by mole	by mass	mol. wt.
AFR	11.329	N/A	N/A
R^2	0.0000%	N/A	N/A
N_2	70.2170%	62.8050%	28.013
O_2	2.9143%	2.9776%	31.999
CO_2	20.8114%	29.2438%	44.010
CO	0.0032%	0.0028%	28.010
H_2O	4.6610%	2.6811%	18.015
Ar	1.1998%	1.5304%	39.948
SO_2	0.1933%	0.3954%	64.059
total	100.0000%	99.6360%	31.205

The adjusted coal composition is shown in this next table:

Table 38. Adjusted Coal Analysis

mol. wt.	by mass	by mole	
12.011	68.86%	48.478%	C
1.008	5.77%	48.410%	H
15.999	4.83%	2.553%	O
14.007	1.24%	0.747%	N
32.060	1.18%	0.310%	S
51.746	11.38%	1.860%	Ash
18.015	6.74%	3.164%	H_2O
8.013	100.00%	105.52%	total

Based on the adjusted coal analysis, we calculate heating values, specific heats, and enthalpies (sensible heats):

Table 39. Additional Coal and Ash Properties

13,322	HHV [BTU/lbm]
12,725	LHV [BTU/lbm]
0.35	Cp coal [BTU/lbm°F]
0.25	Cp ash [BTU/lbm°F]
10.1	SH coal [BTU/lbm]
62.8	SH ash [BTU/lbm]

After making these adjustments to the boiler spreadsheet, we can go on to solve for the flows and efficiencies, as in the first two examples. The steam calculations are:

Table 40. Steam Calculations

499.7	main steam inlet enthalpy [BTU/lbm]
1461.8	main steam exit enthalpy [BTU/lbm]
1314.9	cold reheat enthalpy [BTU/lbm]
1520.4	hot reheat enthalpy [BTU/lbm]
2.749E+09	main steam heat transfer [BTU/hr]
5.212E+08	reheater heat transfer [BTU/hr]
3.270E+09	total heat transfer to steam [BTU/hr]

We calculate the additional parameters as before, using Excel's Solver to adjust the coal mass flow rate to arrive at a zero heat balance error. As this model considers incomplete combustion, the additional heating that would be obtained by completely combusting CO to CO_2 is subtracted from that added by the coal (LHV + sensible heat). This is the mass flow of CO, which is equal to the total flue gas mass flow rate times the mass fraction of CO, times the heat of combustion of carbon monoxide (4342 BTU/lbm or 10,100 kJ/kg). Inclusion of this term complicates the algebraic solution but has no appreciable impact on the Solver algorithm. The additional calculated parameters are:

Table 41. Calculated Parameters

-0.6	ambient air enthalpy [BTU/lbm]
69.0	stack enthalpy [BTU/lbm]
277,993	coal feed rate [lbm/hr]
13,431	ash flow rate [lbm/hr]
3,149,280	inlet air flow rate [lbm/hr]
3,413,842	flue gas flow rate [lbm/hr]
3.537E+09	LHV heat input from coal [BTU/hr]
92.45%	boiler LHV efficiency
88.31%	boiler HHV efficiency

If we calculate the results before adding the CO correction term or multiply it by zero, we find that the efficiency ignoring the incomplete combustion is 92.47% and 88.32% for the LHV and HHV, respectively. This level of CO is typical and so is the magnitude of impact on boiler efficiency. Carbon monoxide is not a significant thermal performance issue; rather, it is an emissions issue. Even for older boilers that lack emission control systems, carbon monoxide does eventually completely oxidize and become CO_2. Toxicity is a near-field concern.

The mass and heat balance results are listed in the following table:

Table 42. Conservation of Mass and Energy

Inputs			
lbm/hr	BTU/hr	fraction	source
3.149E+06	-1.874E+06	-0.023%	from ambient air
2.780E+05	3.540E+09	42.642%	from coal (combustion + sensible)
5.394E+06	4.763E+09	57.381%	from steam
8.821E+06	8.301E+09	100.000%	total
Outputs			
lbm/hr	BTU/hr	fraction	destination
5.394E+06	8.034E+09	96.778%	to steam
3.414E+06	2.356E+08	2.838%	through stack
1.343E+04	8.428E+05	0.010%	with ash
0.000E+00	3.101E+07	0.373%	radiation loss
8.821E+06	8.301E+09	100.000%	total
0.000E+00	0.000E+00		error

Complete SI calculations and results can be found in boiler4.xls

Chapter 7. Mass Leakage

The next aspect of boilers we will consider is mass leakages and there are several. We have already considered heat leakage in the form of radiation and convection to the environment as well as ash streams. As boilers of the type considered here (coal, oil, and gas fired) generally operate on forced air flow, leakages are outward. We consider three types of leaks: 1) after the preheater on the cold side, 2) after combustion and before the heat exchange zone, and 3) before the preheater on the hot side. All of these leakage flow streams carry energy in the form of enthalpy out of the boiler and into the environment without heating the steam and so diminish efficiency.

We must still provide sufficient air for combustion. If some air leaks out on the cold side after the preheater, we must force that much more air into the system to meet combustion requirements, which means more power to the FD fans. [FD fans may be steam or electrically driven. For steam-driven FD fans, the associated loss is within the steam system. For electrically-driven FD fans, the losses are included in auxiliary (station service) load calculations.] None of these flows impact combustion per se by virtue of the way we have separated them or lumped the streams together. This means we can modify the existing spreadsheets to include the leakages and their impact.

We begin with boiler1leak.xls to which we have added the three leakages as a percentage of the total gas flow at that point. These factors (user inputs are blue in the spreadsheet) can be adjusted as desired. The calculations will automatically update. We have also added a radiation/convective UA factor to account for those losses. Instead of specifying the flue gas flow, as we did in boiler1.xls, here we specify the target excess air (18%). Excel's solver is used to adjust the inlet air mass flow rate (cell L24) to match the target excess air (cell L34 must equal A25). As this is not an option with the solver, we create another cell (P18=L34-A25) and require that it be zero.

The ambient conditions in SI units are:

Table 43. Ambient Conditions

0.10	barometric pres. [MPa]
295	ambient dry-bulb temp. [°K]
68%	ambient relative humidity

The design parameters in SI units are listed in this next table:

Table 44. Boiler Design Parameters

1,902,883	main steam flow rate [kg/hr]
30.3	feedwater inlet pres. [MPa]
572	feedwater inlet temp. [°K]
27.3	economizer pres. [MPa]
744	economizer exit temp. [°K]
24.2	main steam exit pres. [MPa]
872	main steam exit temp. [°K]
1,591,788	reheat flow rate [kg/hr]
5.6	reheat inlet pres. [MPa]
639	reheat inlet temp. [°K]
5.2	reheat exit pres. [MPa]
880	reheat exit temp. [°K]
303,942	coal feed rate [kg/hr]
622	air preheat temp. [°K]
422	preheater exit temp. [°K]
474,775	radiation loss UA [kJ/hr/°K]
2%	leak after preheater
2%	leak after combustion
2%	leak before preheater
18%	excess air

The coal analysis and calculated properties (heating values, specific heats, and sensible heats) for the coal and ash are listed in the following table:

Table 45. Coal Analysis and Properties

ox. num.	mol. wt.	by mass	by mole	
1.00	12.011	48.51%	41.141%	C
0.25	1.008	3.25%	32.843%	H
-0.50	15.999	10.69%	6.806%	O
0.00	14.007	0.65%	0.473%	N
1.00	32.060	0.40%	0.127%	S
0.00	51.746	5.50%	1.083%	Ash
0.00	18.015	31.00%	17.528%	H_2O
0.46075	10.186	100.00%	100.00%	total
2.24045	stoichiometric molar air/fuel ratio			
6.32681	stoichiometric mass air/fuel ratio			
19,305	HHV [kJ/kg]			
17,864	LHV [kJ/kg]			
2.16	Cp coal [kJ/kg°K]			
1.05	Cp ash [kJ/kg°K]			
14.4	SH coal [kJ/kg]			
441.9	SH ash [kJ/kg]			

The dry air, ambient humidity, moist air, stoichiometric flue gas, and operating flue gas properties are listed in the table below:

Table 46. Air and Flue Gas Composition and Properties

Composition of Dry Air			
mol. wt.	by mole	by mass	gas
28.013	78.0843%	75.5188%	N_2
31.999	20.9476%	23.1416%	O_2
39.948	0.9367%	1.2919%	Ar
44.010	0.0314%	0.0477%	CO_2
28.965	100.0000%	100.0000%	total

Composition of Moist Air			
0.011565 ambient humidity			
mol. wt.	by mole	by mass	gas
28.013	76.6589%	74.6554%	N_2
31.999	20.5652%	22.8771%	O_2
39.948	0.9196%	1.2771%	Ar
44.010	0.0308%	0.0472%	CO_2
18.015	1.8254%	1.1432%	H_2O
28.76523	100.0000%	100.0000%	total

Stoichiometric Flue Gas			
mol. wt.	by mole	by mass	gas
28.013	67.865%	65.043%	N_2
31.999	0.000%	0.000%	O_2
44.010	16.261%	24.484%	CO_2
18.015	15.010%	9.252%	H_2O
39.948	0.813%	1.111%	Ar
64.059	0.050%	0.110%	SO_2
29.229	100.000%	100.000%	total

Operating Flue Gas			
mol. wt.	by mole	by mass	gas
28.013	69.209%	66.492%	N_2
31.999	3.143%	3.450%	O_2
44.010	13.780%	20.799%	CO_2
18.015	12.995%	8.029%	H_2O
39.948	0.829%	1.136%	Ar
64.059	0.042%	0.093%	SO_2
29.1581	100.000%	100.000%	total

The steam calculations are listed in this next table [note the -97 property formulation for SI units]:

Table 47. Steam Properties and Calculations

-97	steam property formulation
10.0%	economizer pressure drop
11.1%	superheater pressure drop
20.0%	combined pressure drop
8.0%	reheater pressure drop
1322.3	main steam inlet enthalpy [kJ/kg]
3005.2	economizer exit enthalpy [kJ/kg]
3497.5	enthalpy at superheater exit [kJ/kg]
3096.6	cold reheat enthalpy [kJ/kg]
3682.2	hot reheat enthalpy [kJ/kg]
3.202E+09	economizer heat transfer [kJ/hr]
9.368E+08	superheater heat transfer [kJ/hr]
9.322E+08	reheater heat transfer [kJ/hr]
5.071E+09	total heat transfer to steam [kJ/hr]
7.838E+08	heat transferred by preheater [kJ/hr]
5.430E+09	LHV heat input from coal [kJ/hr]
93.40%	boiler LHV efficiency
86.43%	boiler HHV efficiency

Combustion and flue gas calculations are listed in the following table:

Table 48. Combustion and Flue Gas Calculations

439,921	stoichiometric O2 flow rate [kg/hr]
1,922,981	stoichiometric moist air flow rate [kg/hr]
16,717	ash flow rate [kg/hr]
2,315,424	inlet air flow rate [kg/hr]
46,308	leakage after preheater [kg/hr]
51,127	leakage after combustion [kg/hr]
50,104	leakage before preheater [kg/hr]
2,269,115	flow after preheater [kg/hr]
2,556,340	flow before combustion [kg/hr]
2,505,214	flow after combustion [kg/hr]
2,455,109	flue gas (stack) flow [kg/hr]
6.32680	stoichiometric A/F ratio
7.46562	operating A/F ratio

Conservation of mass and energy calculations (as in boiler2.xls, boiler3.xls, and boiler4.xls) has been added to this spreadsheet too so that we can show the leakage streams and how much energy loss they represent. As listed in the table on the right side midway down, the three mass flow leakage rates (2%, 2%, 2%) result in an energy "leakage" of 0.124%, 1.000%, and 0.179%, respectively or an average of 0.434%.

> *The average energy leak is about 1/5th of the average mass leak. This supports the motivation for the Losses Method by reinforcing the argument that estimating some small quantities (losses) can be used to calculate larger ones (mass flow rates or heating values) without introducing excessive additional uncertainty has merit.*

The entering and exiting mass and energy streams are listed in this table:

Table 49. Mass and Energy Streams

Inputs			
kg/hr	kJ/hr	fraction	source
2.315E+06	1.528E+07	0.119%	from ambient air
3.039E+05	5.434E+09	42.141%	from coal (combustion + sensible)
3.495E+06	7.445E+09	57.740%	from steam
6.114E+06	1.289E+10	100.000%	total
Outputs			
kg/hr	kJ/hr	fraction	destination
3.495E+06	1.252E+10	97.069%	to steam
4.631E+04	1.598E+07	0.124%	leak after preheater (cold)
5.113E+04	1.289E+08	1.000%	leak after combustion
5.010E+04	2.310E+07	0.179%	leak before preheater (hot)
2.455E+06	7.105E+06	0.055%	through stack
1.672E+04	7.387E+06	0.057%	with ash
0.000E+00	1.955E+08	1.516%	radiation loss
6.114E+06	1.289E+10	100.000%	total
0.000E+00	0.000%	0.000%	error

In this model we also consider the heat transfers, temperature differences, and conductances, as these are a significant part of any boiler design. The temperatures and enthalpies are listed in the table below:

Table 50. Temperatures and Enthalpies

Point	Temp. [°K]	Enthalpy [kJ/kg]	
1	295.4	6.6	inlet air
2	622.0	345.1	preheated air
3	2230.9	2521.6	after combustion
4	1705.7	1765.5	after superheater
6	710.9	470.0	after economizer
6A	711.0	470.2	after ash removal
7	703.3	461.0	before preheater
8	422.0	141.8	after preheater
A	571.9	1322.3	feedwater
C	744.5	3005.2	after economizer
D	872.0	3497.5	main steam
E	639.0	3096.6	cold reheat steam
F	880.4	3682.2	hot reheat steam

The log-mean-temperature differences, heat transfer, and conductance are listed in this next table:

Table 51. LMTDs and UAs

Points	LMTD	Q	UA	
72/81	102.3	7.84E+08	7.66E+06	preheater
3D/4C	1148.6	9.37E+08	8.16E+05	superheater
3F/4E	1203.0	9.32E+08	7.75E+05	reheater
4C/6A	425.2	3.20E+09	7.53E+06	economizer
-	-	5.85E+09	1.68E+07	total for boiler
60/70	411.8	1.96E+08	4.75E+05	heat loss

The other spreadsheets (boiler2.xls, boiler3.xls, and boiler4.xls) can be similarly be modified to incorporate leakages.

Chapter 8. Sensitivity

In Chapter 4 we considered measurement uncertainty, particularly in the mass flow rates of coal, air, and flue gas plus coal analysis. In Chapter 6 we considered uncertainty in measuring stack gas constituents. Perhaps a bigger question is, "What difference does it make?" How much does the uncertainty in each measured quantity impact the final outcome? Well, we can calculate just that. The mathematical principles we will employ here come from *perturbation theory*. Our motivation is to find out which measurements have the largest impact so that we can focus our efforts (and investment) on those in order to obtain the best practical result.

Let's first consider a few obvious sensitivities. While we haven't specifically stated it this way, it should be obvious that the calculated efficiency of a boiler is directly proportional to the mass flow rate of either the fuel or the flue gas and also the heating value. That is, a 5% change in either flow or heating value results in a 5% change in heat input from the fuel. Heat output to the steam is known much more accurately than heat input and so the associated uncertainty is comparatively negligible. Efficiency is the ratio of the two (heat out/heat in), transferring this 5% change directly to the calculated efficiency.

> Note that a 5% change in efficiency postulated here is not applied ± to the final result, rather this is applied multiplicatively. 5% more than 50% is 52.5%, not 55%. As most boiler efficiencies are approximately 90%, this distinction (multiplicative or incremental change) is not as significant.

In order to compare one sensitivity to another, it is convenient to use a ratio that eliminates the units of the independent parameter. That is, we would rather not be comparing percent per degree to percent per hour to percent per kg or pound. In effect, we are calculating the partial derivative of one calculated quantity (e.g., boiler LHV efficiency) to another (barometric pressure in psia or MPa). Our difference formula for sensitivity becomes:

$$s = \frac{\Delta y}{\left(\frac{\Delta x}{x}\right)} \tag{8.1}$$

We will avoid using enthalpies, which are inherently and unavoidably relative to some reference state, for x. We will use absolute temperatures to avoid this problem (i.e., some real, meaningful values can be zero). By adding 459.67 to x in the bottom of Equation 8.1 when considering a temperature in °F, we obtain the same values of sensitivity, s, using English and SI units, as can be seen by comparing the two tabs in boiler2.xls, which we have modified to perform this sensitivity analysis. To implement this step-by-step differencing and listing of results, we create a button [calculate sensitivities] and assign a macro, as listed below:

```
Option Explicit
Private Sub CommandButton1_Click()
  Dim i As Integer, j As Integer, x As Double, y1 As
     Double, y2 As Double, degr As Double
  j = 24
  For i = 2 To 22
'we loop through rows 2 through 22, skipping 5 and 16,
'which contain headings
    If (i <> 5 And i <> 16) Then
      x = Cells(i, 1).Value
      Cells(i, 1).Value = x * 0.995
      Calculate
      y1 = Range("P18").Value
      Cells(i, 1).Value = x * 1.005
      Calculate
      y2 = Range("P18").Value
      Cells(i, 1).Value = x
      Calculate
      j = j + 1
'we add 459.67 (on the English tab) to the temperatures
      If (i=3 Or i=8 Or i=10 Or i=13 Or i=15) Then
        degr = 459.67
      Else
        degr = 0
      End If
      Cells(j, 1).Value= _
........(y2-y1)/((x*1.005-x*0.995)/(x+degr))
'note: 1/((x*1.005-x*0.995)/x)=100 (see below)
    End If
  Next i
```

The next section is more complicated because when we change one of the coal constituents, we must also adjust the others so that the sum is still 100%.

```
  Dim C As Double, H As Double, O As Double, N As
     Double, S As Double, A As Double, W As Double, z As
     Double
  C = Range("I3").Value
  H = Range("I4").Value
  O = Range("I5").Value
  N = Range("I6").Value
  S = Range("I7").Value
  A = Range("I8").Value
  W = Range("I9").Value
  For i = 1 To 7
    If (i = 1) Then
      x = C * 0.995
      z = (1 - x) / (1 - C)
    ElseIf (i = 2) Then
      x = H * 0.995
```

45

```
            z = (1 - x) / (1 - H)
         End If
   'etc. see VBA for full details
         If (i = 7) Then
            Range("I9").Value = x
         Else
            Range("I9").Value = W * z
         End If
         Calculate
         y1 = Range("P18").Value
         If (i = 1) Then
            x = C * 1.005
            z = (1 - x) / (1 - C)
         ElseIf (i = 2) Then
            x = H * 1.005
            z = (1 - x) / (1 - H)
         ElseIf (i = 3) Then
            x = O * 1.005
            z = (1 - x) / (1 - O)
         ElseIf (i = 4) Then
            x = N * 1.005
            z = (1 - x) / (1 - N)
         ElseIf (i = 5) Then
            x = S * 1.005
            z = (1 - x) / (1 - S)
         ElseIf (i = 6) Then
            x = A * 1.005
            z = (1 - x) / (1 - A)
         End If
         If (i = 1) Then
            Range("I3").Value = x
         Else
            Range("I3").Value = C * z
         End If
   'etc. see VBA for full details
         Calculate
         y2 = Range("P18").Value
         Range("I3").Value = C
         Range("I4").Value = H
         Range("I5").Value = O
         Range("I6").Value = N
         Range("I7").Value = S
         Range("I8").Value = A
         Range("I9").Value = W
         Calculate
         j = j + 1
         Cells(j, 1).Value = 100 * (y2 - y1)
      Next i
   End Sub
```

The results [which are the same for English or SI units and if they weren't, we should be suspicious of the calculations] are listed in order of decreasing significance [conveniently sorted by Excel]:

Table 52. Sensitivities

9.718%	ambient dry-bulb temp.
-9.134%	stack temp.
1.971%	coal C
-0.877%	excess Air
-0.631%	coal H_2O
-0.475%	average surface temp.
-0.432%	feedwater inlet temp.
0.430%	main steam exit temp.
-0.428%	radiation loss UA
0.412%	coal temp.
0.342%	main steam flow rate
0.261%	hot reheat temp.
-0.243%	coal H
-0.222%	cold reheat temp.
0.134%	barometric pres.
-0.134%	ambient relative humidity
-0.105%	ash temp.
0.086%	reheat flow rate
-0.084%	coal Ash
-0.067%	coal O
-0.046%	main steam exit pres.
0.020%	cold reheat pres.
-0.006%	coal N
-0.005%	hot reheat pres.
0.001%	feedwater inlet pres.
0.000%	coal S

These results make sense and also confirm the guidance provided in PTC4 as to the importance of instrumentation and measurements. The ambient temperature, average exiting flue gas temperature at the stack, the coal carbon content, excess air, and coal moisture are the top five items that we must get right in order to accurately measure the efficiency of this type boiler. If this were a gas-fired design, we would see methane in the position where carbon content occupies for coal-fired.

The next nine items are of similar, but lesser importance: average surface temperature (impacting convective and radiative heat loss), feedwater inlet temperature, main steam exit temperature (after the superheater), radiation loss factor (UA), coal temperature (for sensible heat input), main steam flow rate, hot reheat temperature, coal hydrogen content, and cold reheat temperature. The remainder of the variables are of even less significance. It should be noted that is a somewhat low-sulfur coal. Were it a high-sulfur coal, the sensitivity to sulfur content wouldn't be the last item on the list.

Chapter 9. Solutions

Up until this point we have discussed measurement uncertainty and some aspects of calculation uncertainty but each example has been presented in such a way as to arrive at a single unified solution. This is a simplification. The complete picture of boiler testing and calculations is more complicated than this. We will not consider more complex aspects of this problem.

The first example we will consider is adjusting flue gas flow to match excess air. The excess air percentage might be a design or operating criterion or an expectation derived from a measurement during a test. The results are shown in this first figure:

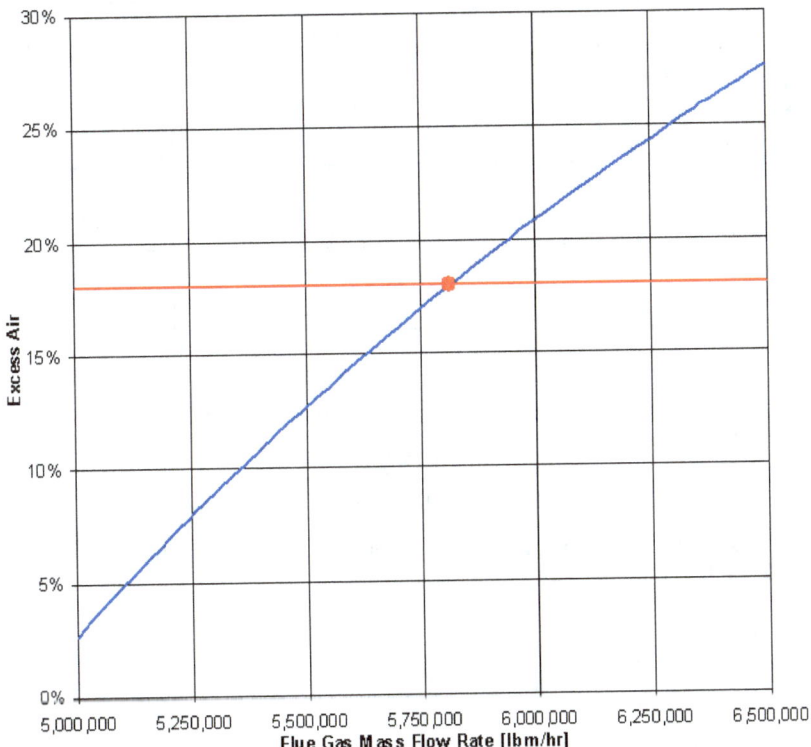

Figure 9. Excess Air vs. Flue Gas Mass Flow Rate

The solution (18% excess air) is easily found (5,814,126 lbm/hr) because this is a simple, almost linear relationship, having only a single meaningful solution. Most any algorithm for solving nonlinear equations in one variable will work.

We next consider adjusting coal mass flow rate to match some desired air/fuel ratio. The results are shown in this second figure:

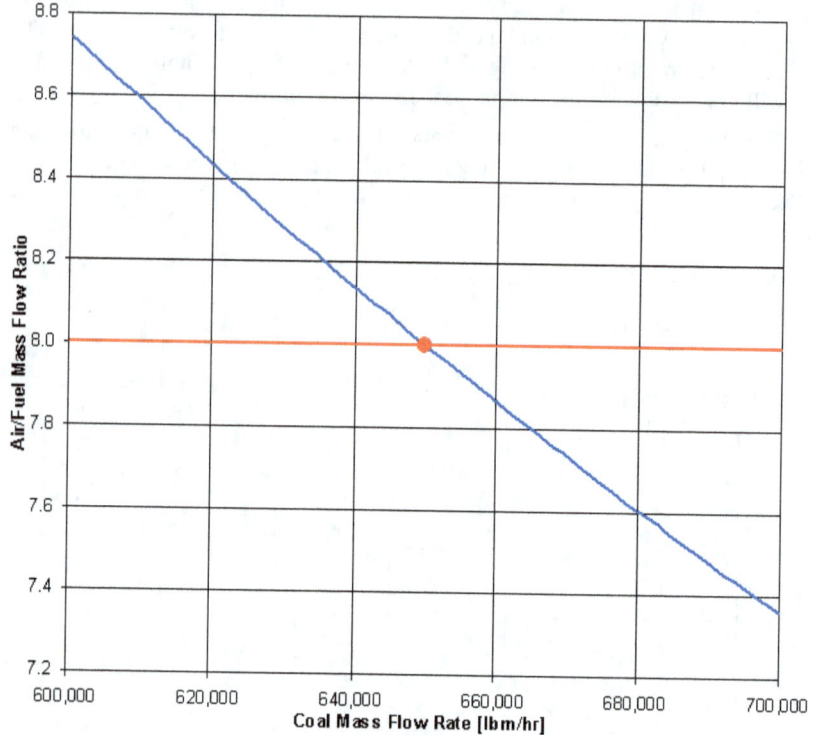

Figure 10. Air/Fuel Ratio vs. Coal Mass Flow Rate

This is also a fairly linear relationship with a single solution (A/F=8 at m_{coal}=649,987 lbm/hr).

The next example is adjusting coal carbon content to match measured flue gas CO_2 mole fraction, which would be reported by the CEMS. This is exactly a linear relationship. The only offset would be the CO_2 in ambient air, which is quite small by comparison.

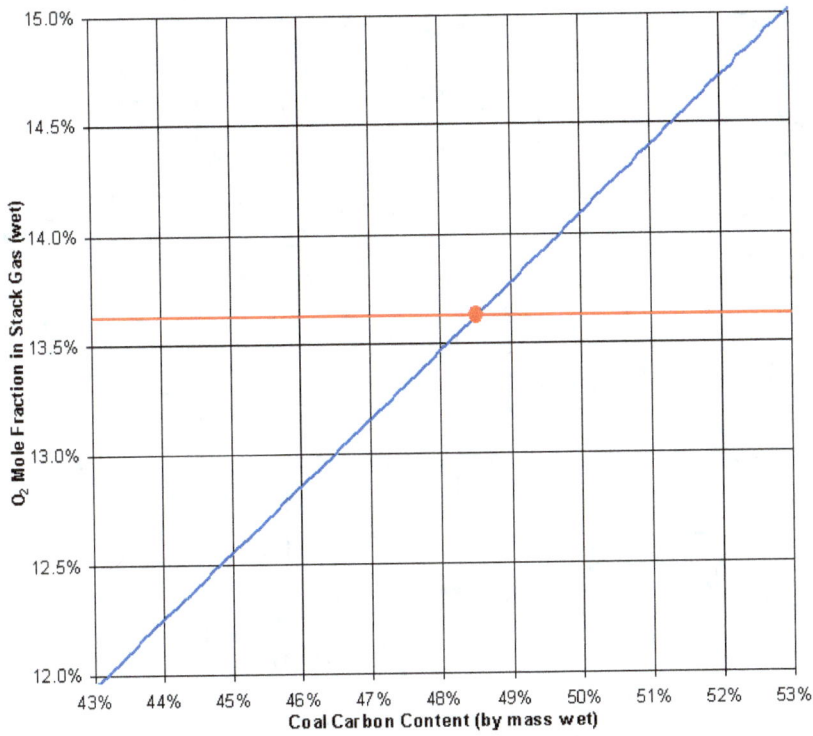

Figure 11. Flue Gas CO_2 vs. Coal Carbon Content

This must be a linear relationship, as coal carbon is the direct source of flue gas CO_2. The solution (48.5162%) is easily found that matches the CEMS measurement (13.625%).

A similar linear relationship exists between coal sulfur content and SO_2 in the stack, as there is even less SO_2 in ambient air.

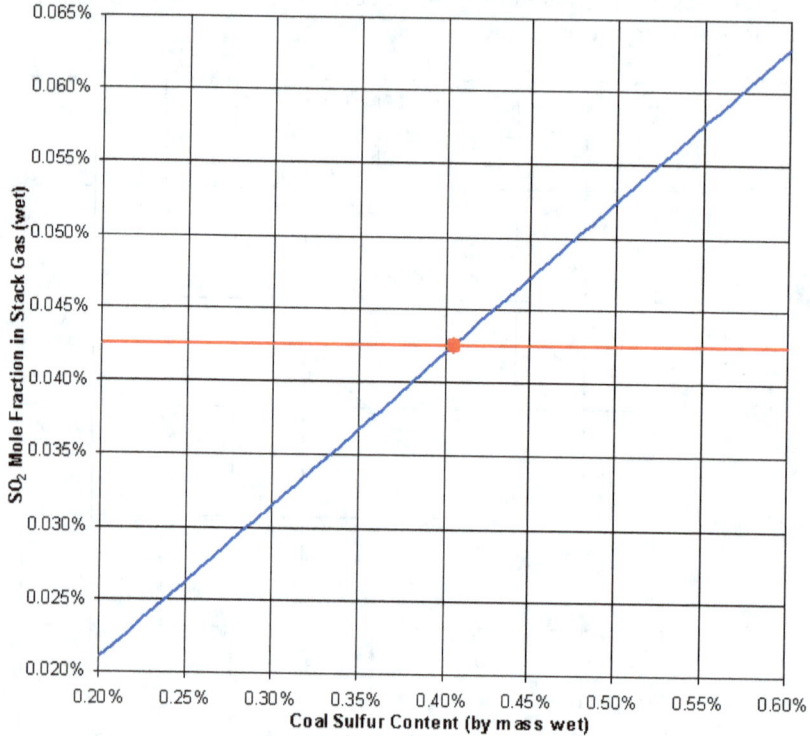

Figure 12. Flue Gas SO_2 vs. Coal Sulfur Content

Again, the solution (0.4048%) is easily found where the calculated SO_2 matches that reported by the CEMS (0.0425%).

These four examples seem simple enough, especially the last two, adjusting the coal composition to match flue gas measurements. Again, this is an incomplete picture; because adjusting the coal carbon content doesn't simply change the flue gas CO_2. It also changes the flue gas SO_2. Likewise, adjusting the coal sulfur content doesn't simply change the flue gas SO_2. It also changes the flue gas CO_2.

We next adjust coal carbon and sulfur content to match flue gas CO_2 and SO_2. First, we create a map showing the residual (combined root sum squared error) for a variety of values. This creates a surface, which is shown in the following figure with colors and contours (blue is small, red is large).

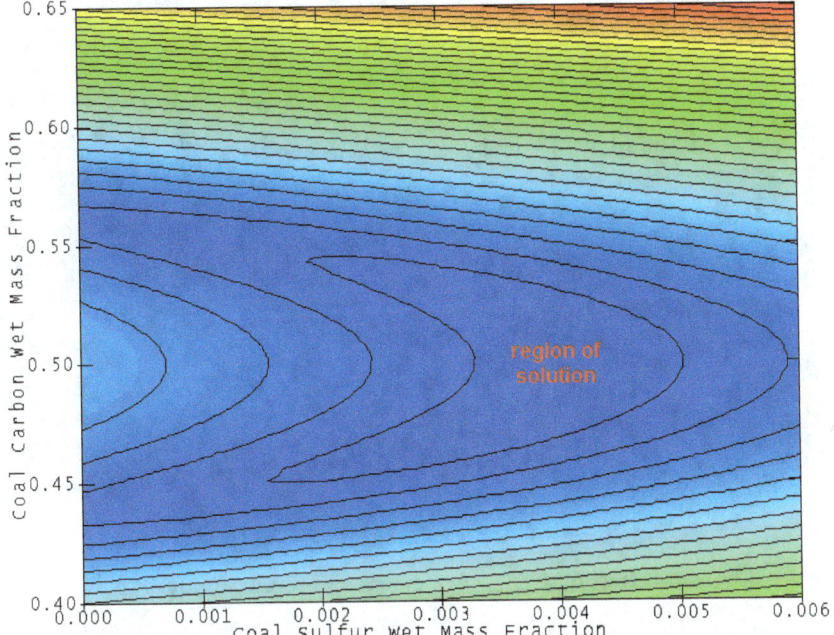

Figure 13. Residual vs. Coal Sulfur and Carbon Content

The best solution lies within the inner-most contour, as indicated by "region of solution" (49.2231% carbon and 0.4107% sulfur). In order to solve this two-dimensional nonlinear problem, we must use more advanced methods. Modified Broyden's method is an excellent choice for this problem (for more details see Appendix I).

This same residual map in 3D is shown in this next figure:

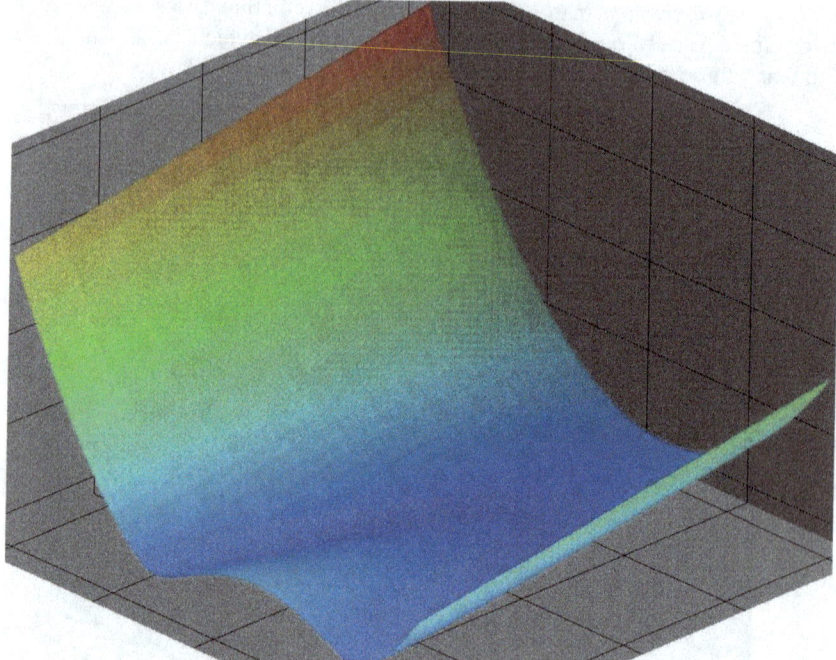

Figure 14. Residual vs. Coal Carbon and Sulfur Content in 3D

The optimal point having the lowest residual (best agreement between calculated CO_2 and SO_2 and that reported by the CEMS) is the bluest area at the bottom of the valley that looks like it has two ravines leading into it from the left.

The last example we consider in this chapter will be simultaneously adjusting coal carbon, sulfur, and moisture content to match values of CO_2, SO_2, and H_2O reported by the CEMS. This is a three-dimensional nonlinear minimization problem easily handled by Broyden's method. It is a challenge to illustrate four variables (carbon, sulfur, moisture, and residual) on a flat surface. The figure below shows layered meshes of constant residual (discrepancy between calculated and measured), colored as before and shown in the legend. The solution lies deep inside the bluest mesh at carbon 32.5093%, sulfur 0.230273%, and moisture 22.9539%.

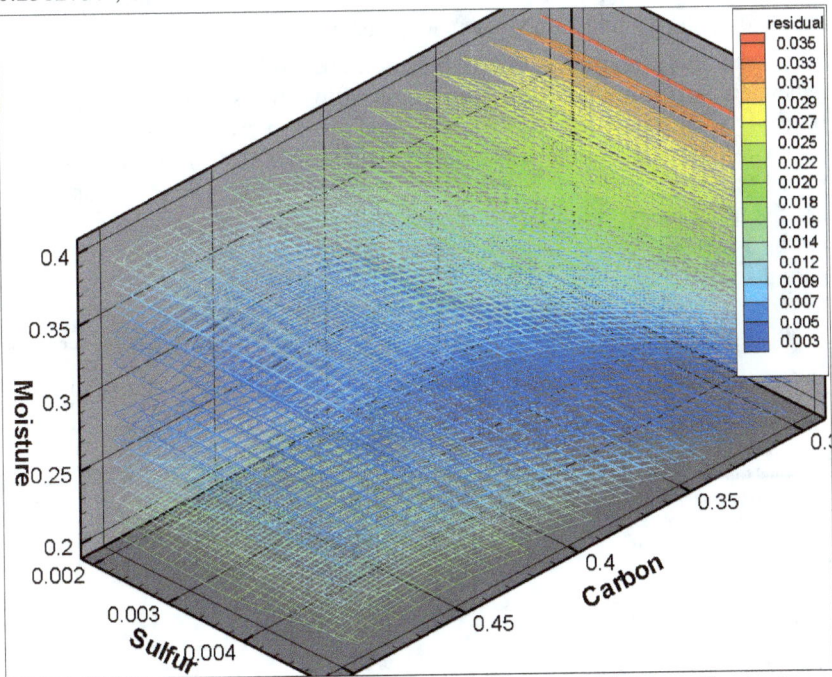

Figure 15. Residual vs. Coal Carbon, Sulfur, and Moisture Content

Chapter 10. Correction Curves

Very few coal-fired boilers come with curves, as manufacturers are reluctant to expose themselves to any more liability than is absolutely necessary. This is not surprising, considering how many things can impact the performance of a coal-fired boilers and no manufacturer wants to be embroiled in endless squabbles, especially with customers who may not appreciate all of the subtle nuances. While I do have such curves, these are proprietary and not available for open disclosure. Even the common redactions would still leave enough information to identify the source. This is the case for most older boilers supplied by the manufacturers with a long history.

Some more recent coal-fired boilers come with curves that have been generated by perturbing a PTC4 model, tuned to the design conditions. This is basically what we did in Chapter 8 at the bottom of spreadsheet boiler2.xls, only for more inputs. There are many subtle inputs to a PTC4 model. A simplified model (still containing the core calculations) may be found in PTC4.xls. There are the usual inputs much like the previous boiler?.xls spreadsheets and also a button at the top plus a series of tables and figures based on those tables. A partial screenshot is shown below:

	A	B	C	D	E	F	G	H	I
1	PTC4 SIMPLIFIED CALCULATIONS			Tabulated Values Calculated by Perturbing					
2		calculate		Pamb	Corr.	RH	30	40	50
3	**Ambient Conditions**	Value	Units	12.50	0.9999	0%	1.0003	1.0004	1.0005
4	ambient pressure	14.7	psia	12.79					
5	ambient dry-bulb	60.0	Deg F	13.07					
6	ambient relative humidity	60%	%	13.36					
7	**Coal Composition**			13.64					
8	carbon mass fraction	48.50%	%	13.93					
9	hydrogen mass fraction	3.25%	%	14.21					
10	oxygen mass fraction	10.69%	%	14.50					
11	nitrogen mass fraction	0.65%	%	14.79					
12	sulfur mass fraction	0.40%	%	15.07					
13	ash coal mass fraction	5.50%	%	15.36					
14	moisture (H$_2$O)	31.00%	%	15.64					
15	total	99.99%	%	15.93					
16	**Normalized Coal Composition**			16.21					
17	carbon mass fraction	48.50%	%	16.50					
18	hydrogen mass fraction	3.25%	%						
19	oxygen mass fraction	10.69%	%	Tdb					
20	nitrogen mass fraction	0.65%	%	30					
21	sulfur mass fraction	0.40%	%	35					
22	ash coal mass fraction	5.50%	%	40					
23	moisture (H$_2$O)	31.00%	%	45					
24	total	100.00%	%	50					
25	**Ash Composition**			55					
26	fly ash	76.92%	%	60					
27	bottom ash	15.38%	%	65					
28	other	7.69%		70					

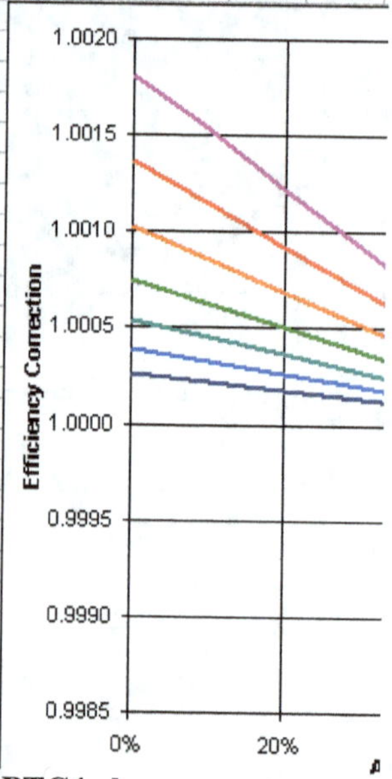

Figure 16. Partial Screenshot of PTC4.xls

The [calculate] button is assigned to a macro that varies one or more inputs, recalculates the boiler efficiency, and copies the results into the corresponding table on the right beneath the curves, which display the calculated efficiency adjustments. An efficiency adjustment of 1.05 would be 85.73%×1.05=90.01%. The VBA procedure for the button is listed in part below (see spreadsheet for complete listing):

```
'vary ambient pressure

irow = 2
x = Range("B4").Value 'save value
For i = 1 To 15
  Range("B4").Value = 12.5 + (i - 1) * 4 / 14
  Calculate
  XX(i) = Range("B4").Value
  YY(i, 1) = Range("B48").Value
Next i
For i = 1 To 15
  irow = irow + 1
  Cells(irow, 4).Select
  Cells(irow, 4).Value = XX(i)
  Cells(irow, 5).Value = YY(i, 1) / YY(8, 1)
Next i
Range("B4").Value = x 'restore original value

'vary ambient temperature

irow = irow + 2
x = Range("B5").Value 'save value
For i = 1 To 15
  Range("B5").Value = 25 + 5 * i
  Calculate
  XX(i) = Range("B5").Value
  YY(i, 1) = Range("B48").Value
Next i
For i = 1 To 15
  irow = irow + 1
  Cells(irow, 4).Select
  Cells(irow, 4).Value = XX(i)
  Cells(irow, 5).Value = YY(i, 1) / YY(8, 1)
Next i
Range("B5").Value = x 'restore original value
```

Note that the initial value for each perturbed parameter is first saved, then varied in a loop, and finally restored. This spreadsheet uses several sections of code that also appear in boiler?.xls (psychrometrics, flue gas enthalpy, and basic water properties) and also contains the simplified PTC4 calculations, as listed below:

```
Function Boiler(H2O As Double, C As Double, H2
    As Double, N As Double, S As Double, Ash As
    Double, O As Double, PerFly As Double, PerBot
    As Double, FlyComb As Double, BotComb As
    Double, AHlk As Double, Pamb As Double, Tamb
    As Double, RH As Double, O2dry As Double, CO
    As Double, AHGout As Double, AHAin As Double,
    FuelFlow As Double, FuelHV As Double) As Variant
  Dim iter As Integer, CO2 As Double,
    CO2lb As Double, CO2min As Double, CO2max As
    Double, COlb As Double, ExcessAir As Double,
    HC As Double, L As Double, Lbeta As Double,
    lbFactor As Double, Lgp As Double, Lh As
    Double, Lma As Double, Lmf As Double, Luc As
    Double, maxCO2 As Double, Moisture As Double,
    Moisturelb As Double, N2 As Double, N2lb As
    Double, NewCO2 As Double, O2lb As Double,
    SpecHum As Double, Sumlb As Double, SumlbMoist
    As Double, Sumlbr As Double, tg15nl As Double,
    wap As Double, wgp As Double, wgpN2 As Double
  SpecHum=fWrh(Pamb,Tamb,RH)
  maxCO2=(31.3*C+11.5*S)/(1.504*C+3.55*H2+0.56*S
    +0.13*N-0.45*O)/100
  CO2min=0
  CO2max=maxCO2
  For iter = 1 To 32
    CO2=(CO2min+CO2max)/2
    N2=1-CO2-O2dry
    If(N2<=0)Then
      maxCO2=CO2
    Else
      ExcessAir=(O2dry-CO/2)/(0.2682*N2-(O2dry-CO/2))
      wgp=(44.01*CO2+32*O2dry+28.02*N2+28.01*CO)
        /(12.01*(CO2+CO))*(C+12.01/32.09*S)
      wgpN2=wgp*28.02*N2/((44.01*CO2+32*O2dry
        +28.02*N2+28.01*CO))
      wap=(wgpN2-N)/0.7685
      Moisture=H2O+8.936*H2+wap*SpecHum
      NewCO2=maxCO2/(1+ExcessAir)
      If(NewCO2>CO2)Then
        CO2min=CO2
      Else
        CO2max=CO2
      End If
    End If
  Next iter
  tg15nl=AHGout+AHlk*(AHGout-AHAin)*0.24/0.26
  Luc=14386*(Ash/(1-(PerFly*FlyComb+PerBot*BotComb))
    -Ash)/FuelHV
```

```
Lgp=wgp*(hent(CO2,0,O2dry,N2,0,0,tg15nl)
  -hent(CO2,0,O2dry,N2,0,0,Tamb))/FuelHV
Lmf=H2O*(hptv(tg15nl)-hptl(Tamb))/FuelHV
Lh=H2*8.936*(hptv(tg15nl)-hptl(Tamb))/FuelHV
Lma=wap*SpecHum*(hptv(tg15nl)-hptv(Tamb))/FuelHV
Lbeta=0.0025
L=Luc+Lgp+Lmf+Lh+Lma+Lbeta
HC=wap*(0.24*(Tamb-80))+0.3*(Tamb-80)
Boiler=Array(ExcessAir,Lma,Lh,Lmf,Lgp,L,1-L)
End Function
```

The CO_2 must be iteratively corrected (see loop on "iter"). A bisection search is used, which always arrives at a reasonable result and never becomes unstable, as is the case with the Newton-Raphson and Regula-Falsi methods. The table for the impact of coal ash on boiler HHV efficiency for these particular inputs is listed in the following table. The other tables can be found in the spreadsheet.

Table 53. Boiler Efficiency Corrections for Coal Ash Content

Ash	5,000	6,000	7,000	8,000	9,000	10,000	11,000
2.8%	0.9919	0.9935	0.9947	0.9954	0.9960	0.9965	0.9968
3.1%	0.9925	0.9941	0.9951	0.9958	0.9964	0.9968	0.9971
3.3%	0.9934	0.9948	0.9957	0.9963	0.9968	0.9971	0.9974
3.7%	0.9944	0.9956	0.9963	0.9969	0.9973	0.9976	0.9978
4.0%	0.9956	0.9965	0.9971	0.9975	0.9978	0.9981	0.9983
4.5%	0.9969	0.9975	0.9980	0.9983	0.9985	0.9987	0.9988
5.0%	0.9984	0.9987	0.9989	0.9991	0.9992	0.9993	0.9994
5.5%	1.0000	1.0000	1.0000	1.0000	1.0000	1.0000	1.0000
6.1%	1.0018	1.0014	1.0012	1.0010	1.0009	1.0008	1.0007
6.7%	1.0037	1.0029	1.0024	1.0021	1.0018	1.0016	1.0014
7.4%	1.0057	1.0045	1.0037	1.0032	1.0028	1.0025	1.0022
8.1%	1.0079	1.0062	1.0052	1.0044	1.0038	1.0034	1.0031
8.8%	1.0101	1.0080	1.0067	1.0057	1.0050	1.0044	1.0039
9.6%	1.0125	1.0099	1.0082	1.0070	1.0061	1.0054	1.0049
10.4%	1.0150	1.0119	1.0099	1.0084	1.0073	1.0065	1.0058

Note that the values in this and the other tables are equal to the calculated efficiency at the values indicated (e.g., 2.8% ash and 5,000 BTU/lbm HHV in the top left corner) divided by the efficiency at the corresponding reference conditions (in this case 5.5% ash and 8,300 BTU/lbm). Note that the value in the tables at the reference conditions is always 1.0000 (i.e., no correction). You would multiply by this number to calculate the expected efficiency. If you wanted to adjust boiler test results back to a reference condition, you would divide by this number.

The expected impact of barometric pressure on this boiler at these reference conditions is shown in this first graph:

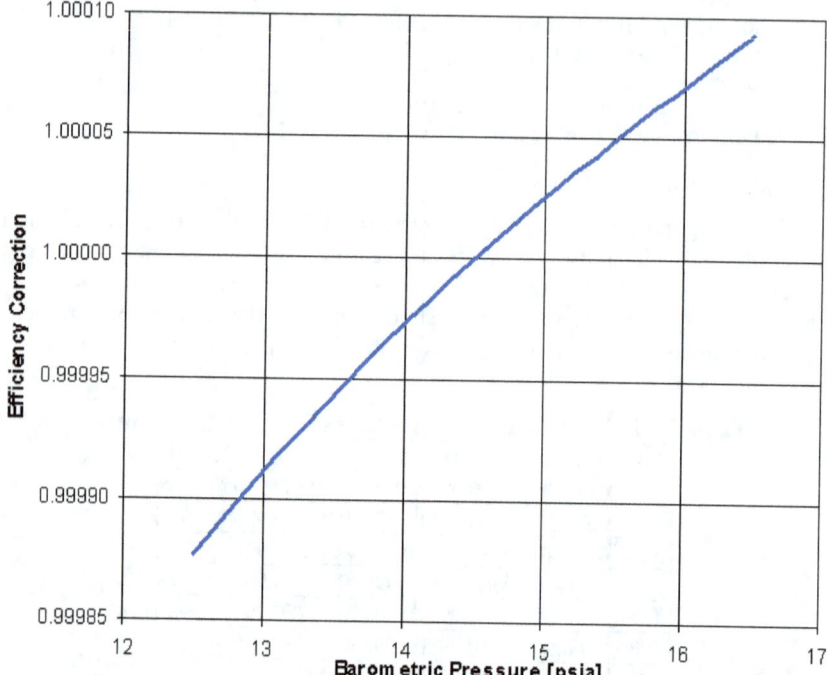

Figure 17. Impact of Barometric Pressure on Boiler HHV Efficiency

As we might expect, increasing barometric pressure increases boiler efficiency. This makes sense because the air and flue gas are more dense, which also results in higher enthalpy per unit volume and also higher thermal conductivity. The volume and surface area of the boiler remain constant but the various processes are more efficient.

The impact of ambient dry-bulb temperature on boiler efficiency is shown in this next figure:

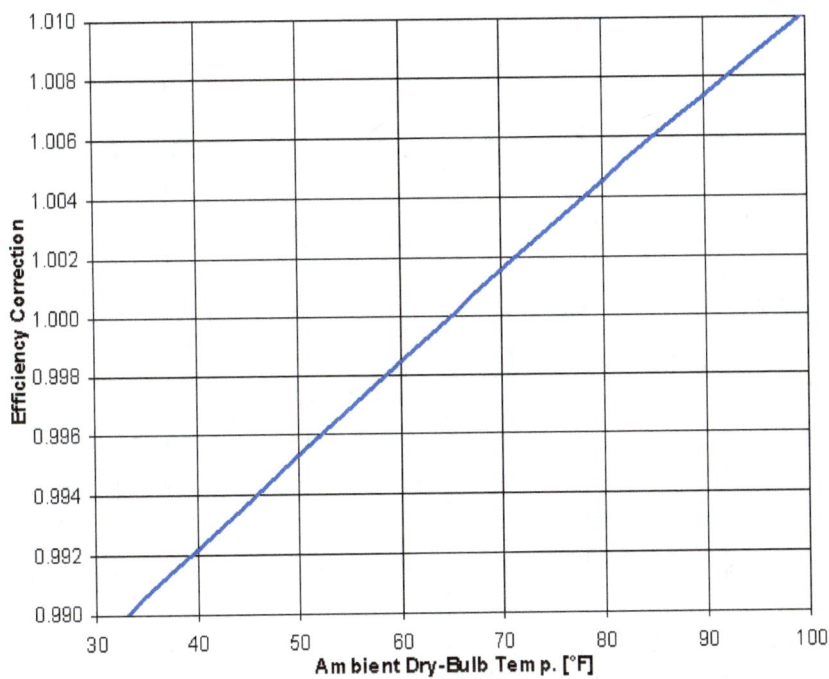

Figure 18. Impact of Ambient Dry-Bulb Temperature on Boiler Efficiency

It is not surprising that hotter ambient temperatures improve boiler efficiency, as this inherently reduces losses to the environment.

When considering ambient relative humidity, it is necessary to also consider ambient dry-bulb temperature; therefore, we build a two-dimensional table to illustrate this relationship.

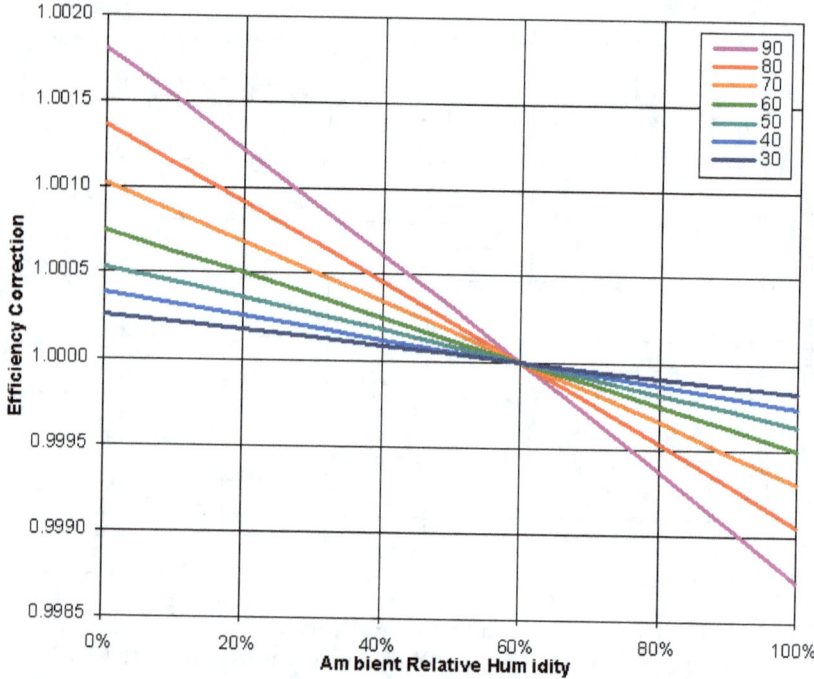

Figure 19. Impact of Ambient Relative Humidity on Boiler Efficiency

In this figure we see that increasing ambient humidity reduces boiler efficiency for all temperatures. This is also not surprising, as this water vapor does not assist or contribute to combustion and carries more energy away into the environment, which is not used to heat the steam.

We next consider the impact of coal moisture content. As with relative humidity, we must simultaneously consider the heating value of the coal so that again we get a two-dimensional table and multiple curves.

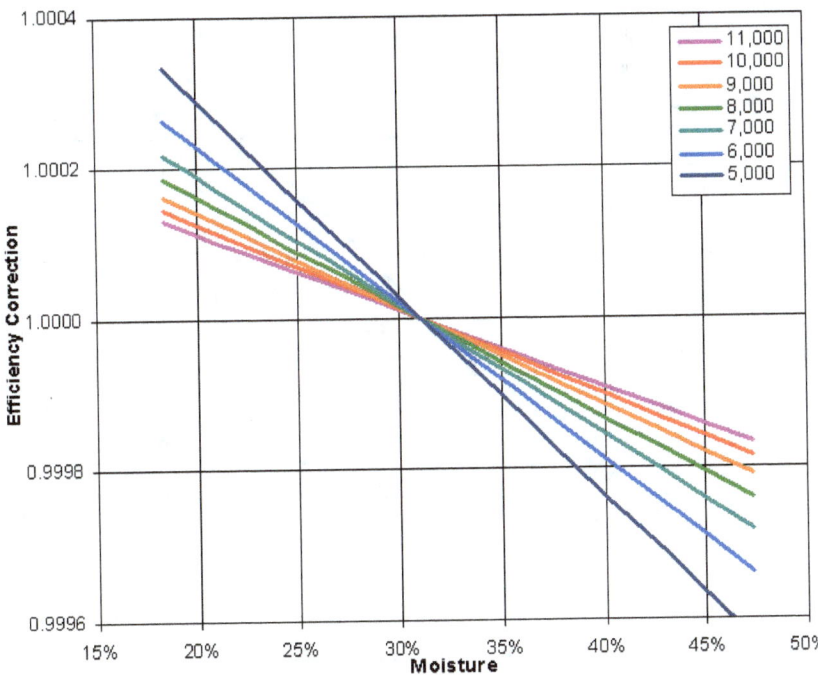

Figure 20. Impact of Coal Moisture on Boiler Efficiency

Here we see that increasing coal moisture consistently leads to lower boiler efficiency. This is for much the same reason as ambient relative humidity, as this water (not to be confused with the hydrogen available for combustion) does not contribute to the process; rather, it carries more energy through the stack and out into the environment not heating the steam.

We next consider the impact of coal ash content. As with relative humidity, we must simultaneously consider the heating value of the coal so that again we get a two-dimensional table and multiple curves.

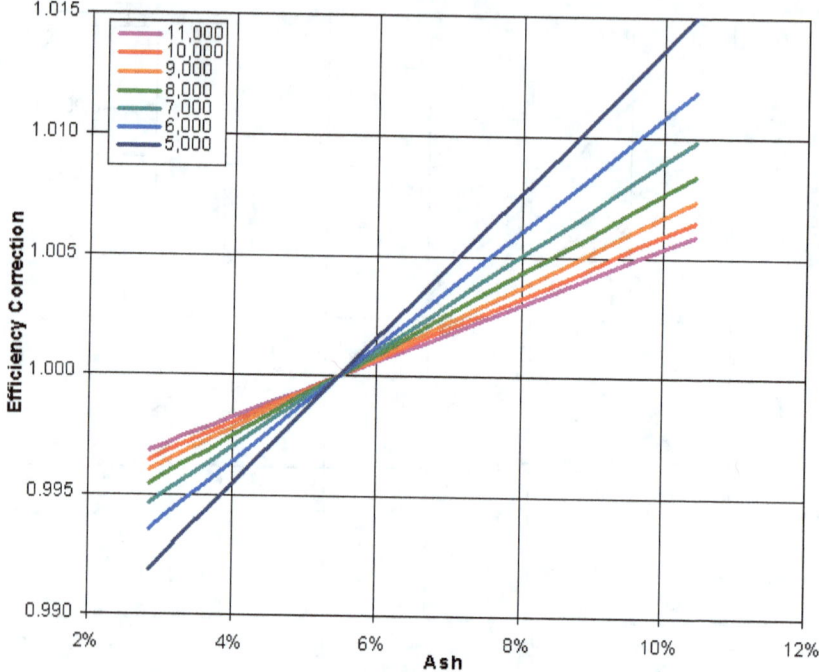

Figure 21. Impact of Coal Ash on Boiler Efficiency

This figure shows that boiler efficiency increases with coal ash content. This is somewhat misleading. Notice that the fuel HHV (see separate curves and legend in upper left corner) is held constant for these parametric runs. In order for the ash to increase and the total to remain 100%, the moisture would have to decrease, as would the carbon and hydrogen, which would effectively reduce the heating value, were we calculating it rather than holding it constant. There are countless scenarios for how we might vary ash and what we might hold constant or how we might compensate accordingly but those exercises are beyond the scope of this text. If you had a particular scenario in mind, you could easily modify the spreadsheet (PTC4.xls) to implement it and see the results.

We next consider the impact of heating value on boiler efficiency. To calculate this, we hold everything else constant and vary HHV. Of course, this isn't physically possible. The heating value arises from the composition so that these factors are not independent. Still, it is illustrative to consider.

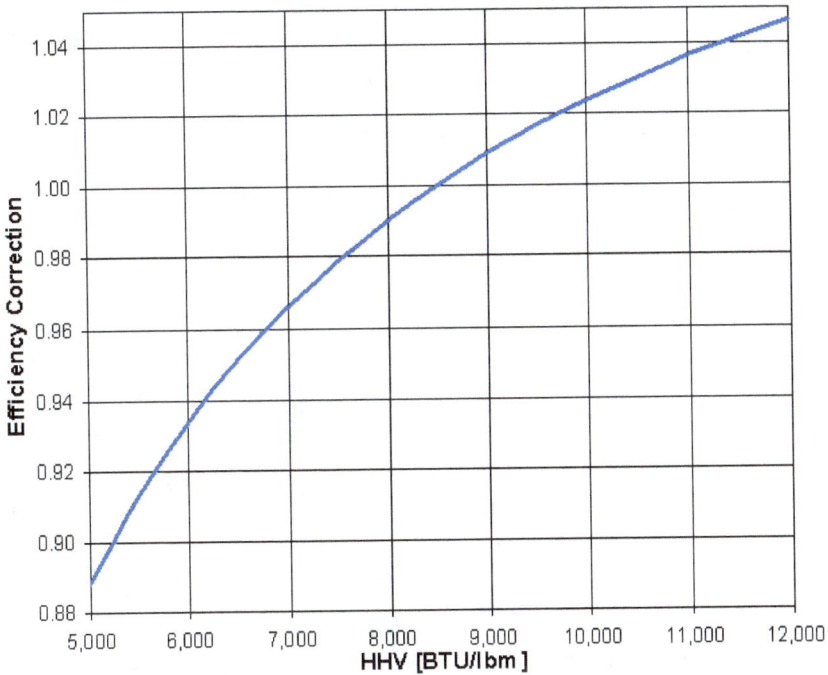

Figure 22. Impact of Heating Value on Boiler Efficiency

Not surprisingly, efficiency increases with heating value. If for no other reason, this is motivation enough to obtain the best coal available for the price. Obviously, high-carbon fuel will be more efficient than burning peat moss or leaves. Notice also that the scale on this last figure is much larger than the previous ones, indicating that the quality (heating value) of the coal is more important than lesser variables.

The last parameter we will consider in this chapter is the impact of C/H ratio; that is, the carbon-to-hydrogen content. Lower C/H means more carbon and less hydrogen. Higher C/H means more hydrogen and less carbon.

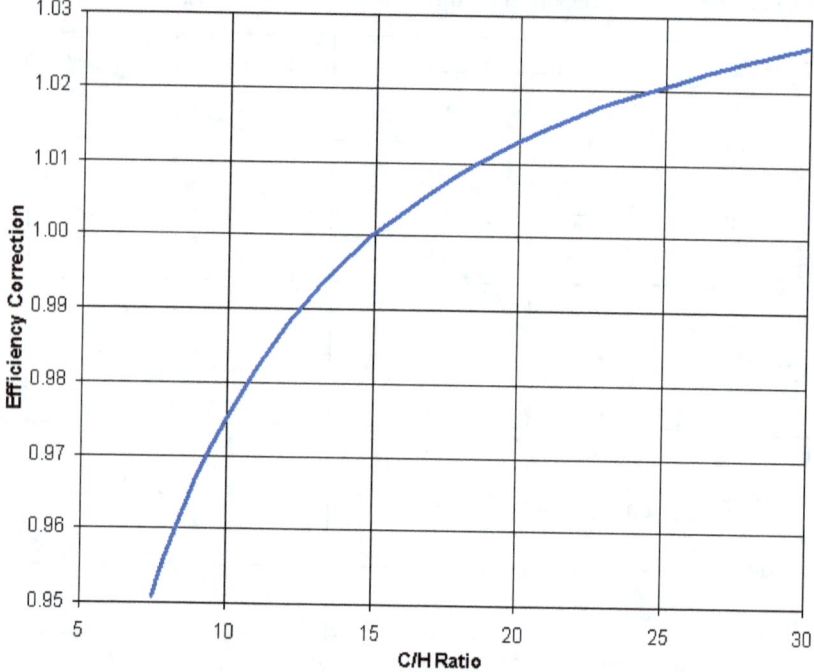

Figure 23. Impact of C/H Ratio on Boiler Efficiency

This figure shows that efficiency increases with increasing C/H ratio or with more carbon and less hydrogen. This may seem counterintuitive, considering the heating value of hydrogen is 8½ times that of carbon per unit mass. As with the previous curves, this one is calculated while holding heating value constant; so there is some room for confusion or ambiguity, as with the previous two curves. The real culprit here is the formation of H_2O vs. CO_2. Water vapor carries considerably more energy than carbon dioxide, not only due to the latent heat of vaporization but also due to almost twice the specific heat. Exhausting more hot H_2O and less hot CO_2 means more unrecoverable heat lost to the environment and consequently lower boiler efficiency.

Chapter 11. Testing

In this chapter we will consider a typical coal-fired boiler test, which is quite complicated and involves many instruments and calculations. This particular plant doesn't have a calibrated flow nozzle on the feedwater as it enters the boiler or the cold reheat steam so that these flow rates must be calculated from other measurements, which means a heat balance around most of the feedwater heaters.

There are a total of 96 measurements that directly or indirectly impact the calculated boiler efficiency. These include 3 ambient conditions; 17 coal-related analyses, measurements, or assumptions; 4 loss parameters related to the ash (both fly and bottom ash); 5 air preheater measurements or estimates; 4 flue gas measurements; 1 power measurement; and 62 steam-related measurements. All of the information and calculations that go into a PTC4 test make a very large spreadsheet. We will consider only the summary inputs and final results that make up the uncertainty analysis. You will find these details in the spreadsheet uncertainty.xls, the top of which is shown below:

	A	B	C	D	E	F
1	**COAL-FIRED BOILER EFFICIENCY TEST UNCERTAINTY**					
2				Measurement Uncer		
3		Test Value		Systematic		
4	TEST UNCERTAINTY (Absolute Basis) (95% Confidence Level)	Mean, $\overline{X}$	Units	B_{inst} Instrument Systematic Uncertainty	$B_{spatial}$ Spatial Systematic Uncertainty	$U_{95,SYS}$ Overall Systematic Uncertainty
6	**Ambient Data**					
7	Barometric Pres.	14.59	psia	3.25E-03	0.00E+00	3.25E-03
8	Ambient Relative Humidity	57.3	%	2.00E+00	0.00E+00	2.00E+00
9	Ambient Dry Bulb Temp.	49	°F	2.60E-01	0.00E+00	2.60E-01
10	**Proximate Fuel Analysis**					
11	Moisture	42.24	wt %	9.16E-01	0.00E+00	9.16E-01
12	Volatile Matter	31.26	wt %	7.05E-01	0.00E+00	7.05E-01
13	Fixed Carbon	6.80	wt %	1.45E-01	0.00E+00	1.45E-01
14	Ash	19.71	wt %	1.97E+00	0.00E+00	1.97E+00
15	**Ultimate Coal Analysis**					
16	Moisture	31.00	wt %	6.25E-01	0.00E+00	6.25E-01
17	Carbon	48.50	wt %	1.12E+00	0.00E+00	1.12E+00
18	Hydrogen	3.25	wt %	6.52E-02	0.00E+00	6.52E-02
19	Nitrogen	0.65	wt %	1.30E-02	0.00E+00	1.30E-02
20	Sulfur	0.40	wt %	8.25E-03	0.00E+00	8.25E-03
21	Ash	5.50	wt %	1.13E-01	0.00E+00	1.13E-01
22	Oxygen	10.69	wt %	2.17E-01	0.00E+00	2.17E-01
23	High Heating Value, HHV	8,371	BTU/lbm	7.00E+01	0.00E+00	7.00E+01

Figure 24. Coal-Fired Uncertainty Spreadsheet (top left)

While the specifics of boiler testing are covered in PTC4, details of the uncertainty analysis are covered in PTC19.1[6]. We arrange the spreadsheet this

---

[6] Performance Test Code 19.1: Test Uncertainty, American Society of Mechanical Engineers, 2005.

way to facilitate the uncertainty analysis. On the left are all of the measurements and other inputs. Some are design parameters and others are assumptions or estimates, as in the case of leakages, which would be even more difficult to measure than actual flows. The second column contains the average (mean) over the test period (usually one hour). The third column lists the unit. The fourth column contains the systematic uncertainty (i.e., measurement bias). The fifth column contains the spatial uncertainty (we might have four or more temperature sensors arranged in the stack to detect variation across the flow). The sixth column contains the total systematic uncertainty (i.e., measurement bias), which is the root-sum-square of the instrument and spatial contributions.

$$U_{systematic} = \sqrt{B_{instrument}^2 + B_{spatial}^2} \qquad (11.1)$$

G	H	I	J	K	L	M	N	P
- REFER TO PTC4 (BOILERS) AND PTC19.1 (UNCERTAINTY)								
rtainty Budget				Uncertainty of Test Results				
	Random		Total	Corrected Boiler Efficiency			89.07	part
$S_{\bar{x}}$	$t_{95,\nu}$	$U_{95, RAND}$	$U_{95, TOT}$	θ	$U_{P1,SYS}$ [kW]	$U_{P1, RAND}$ [kW]	U_{P1} [kW]	Contribution
Standard Deviation of the Mean	Student's t	Random Uncertainty	Total Measurement Uncertainty	Absolute Sensitivity	Systematic Uncertainty of Corrected Power	Random Uncertainty of Corrected Power	Total Uncertainty of Corrected Power	to total uncertainty
1.29E-03	1.96E+00	2.54E-03	4.12E-03	-4.93E-05	-1.60E-07	-1.25E-07	2.03E-07	0.0104%
3.13E-01	1.96E+00	6.16E-01	2.09E+00	1.27E-05	2.54E-05	7.83E-06	2.66E-05	0.1193%
1.35E-01	1.96E+00	2.64E-01	3.71E-01	5.67E-03	1.47E-03	1.50E-03	2.10E-03	1.0612%
2.87E-01	1.27E+01	3.65E+00	3.76E+00	1.60E-03	1.47E-03	5.84E-03	6.02E-03	1.7958%
1.83E-01	1.27E+01	2.33E+00	2.43E+00	6.91E-04	4.87E-04	1.61E-03	1.68E-03	0.9485%
2.79E-01	1.27E+01	3.55E+00	3.55E+00	3.06E-04	4.43E-05	1.09E-03	1.09E-03	0.7632%
1.72E-01	1.27E+01	2.18E+00	2.94E+00	2.52E-04	4.97E-04	5.51E-04	7.42E-04	0.6305%
2.87E-01	1.27E+01	3.65E+00	3.70E+00	-6.17E-02	-3.85E-02	-2.25E-01	2.28E-01	11.0582%
1.34E-01	1.27E+01	1.71E+00	2.04E+00	-5.55E-02	-6.23E-02	-9.47E-02	1.13E-01	7.7918%
1.99E-02	1.27E+01	2.53E-01	2.61E-01	1.38E-03	8.99E-05	3.49E-04	3.60E-04	0.4392%
6.70E-03	1.27E+01	8.52E-02	8.62E-02	-9.69E-04	-1.26E-05	-8.25E-05	8.35E-05	0.2115%
9.72E-03	1.27E+01	1.23E-01	1.24E-01	-2.08E-02	-1.71E-04	-2.57E-03	2.57E-03	1.1735%
1.83E-01	1.27E+01	2.33E+00	2.33E+00	-2.16E-02	-2.45E-03	-5.04E-02	5.04E-02	5.1969%
1.46E-01	1.27E+01	1.85E+00	1.86E+00	2.00E-02	4.33E-03	3.70E-02	3.72E-02	4.4658%
2.67E+01	1.27E+01	3.39E+02	3.46E+02	3.59E-04	2.51E-02	1.22E-01	1.24E-01	8.1567%

Figure 25. Coal-Fired Uncertainty Spreadsheet (top right)

The seventh column (S_X) is the standard deviation of the mean (i.e., the temporal variability of the instrument reading). The eighth column is Student's t, which scales the preceding variability based on the number of measurements.

> Student's t is beyond the scope of this book. The reader is directed to the many publications on statistics that are readily available. The t-value ranges from 1.96 for an infinite number of measurements to 12.7 for a single measurement and is an estimate of what we don't know based on the number of measurements if the process were purely random and normally distributed.

The ninth column is the random uncertainty, as estimated by the standard deviation times Student's t.

$$U_{random} = \sigma t \qquad (11.2)$$

The tenth column is the total uncertainty (in this particular item), which is equal to the root-sum-square of the systematic and random components.

$$U_{total} = \sqrt{U_{systematic}^2 + U_{random}^2} \qquad (11.3)$$

The eleventh column (θ) is the sensitivity, which is the partial derivative of the result (in this case calculated boiler HHV efficiency) with respect to the variable in question.

$$\theta_i = \frac{\partial y}{\partial x_i} \qquad (11.4)$$

While some of these partial derivatives are obvious or can be calculated analytically, most cannot or it is not practical to do so. This is another reason why we use a spreadsheet to perform this analysis. We simply create a button that works much like the one in the previous chapter that perturbs each of the inputs and calculates the change in the result. It's a little more complicated than this, especially when the inputs vary over orders of magnitude, but the same basic approach is used.

The next three columns are the sensitivity times the total systematic contribution for this particular input, the sensitivity times the random component, and the root-sum-square of these two.

$$U_i = \sqrt{(\theta U_{systematic})^2 + (\theta U_{random})^2} \qquad (11.5)$$

The contribution to the total uncertainty for this particular input is shown in the far right column as a percentage of the total uncertainty, not the change in efficiency itself. We must accumulate these calculations for every input that contributes to the final result (96 in all for this test of this boiler, which will be different for other cases).

The top and bottom right corners of the table are shown in this next figure (see uncertainty.xls for complete details).

	K	L	M	N	P
	\multicolumn{5}{l	}{**ND PTC19.1 (UNCERTAINTY)**}
	\multicolumn{5}{l	}{Uncertainty of Test Results}
	\multicolumn{3}{l	}{Corrected Boiler Efficiency}	89.07	part
	θ	$U_{P1,SYS}$ [kW] Systematic Uncertainty of Corrected Power	$U_{P1,RAND}$ [kW] Random Uncertainty of Corrected Power	U_{P1} [kW] Total Uncertainty of Corrected Power	Contribution to total uncertainty
	Absolute Sensitivity				
	-3.44E-07	-1.72E-07	-4.07E-08	1.77E-07	0.0097%
	-3.44E-07	-1.72E-07	-1.46E-07	2.26E-07	0.0110%
	1.58E-05	7.88E-06	2.91E-06	8.40E-06	0.0671%
	1.58E-05	7.88E-06	6.27E-06	1.01E-05	0.0735%
	-2.77E-07	-1.38E-07	-5.24E-08	1.48E-07	0.0089%
	-2.77E-07	-1.38E-07	-5.24E-08	1.48E-07	0.0089%
	7.37E-06	3.68E-06	1.40E-06	3.94E-06	0.0459%
	RSS	0.122	0.554	0.57	100.00%

Figure 26. Coal-Fired Uncertainty Spreadsheet (bottom right corner)

The calculated HHV boiler efficiency for this test period is shown in the upper right corner (89.07%). The total uncertainty is 0.57% (bottom right). This means that there is a 95% certainty that the actual efficiency is somewhere between 89.07%±0.57% or from 88.51% to 89.64%.

> *The 95% confidence interval and how that must be considered in the calculations in order to arrive at this result are beyond the scope of this text. The reader is directed to PTC19.1 for more details on this matter.*

Let is now consider the top twelve contributors:

Table 54. Top 12 Uncertainty Contributions

Measurement	contribution	cumulative
Fly Ash Unburned Carbon	16.0283%	16.0283%
Moisture	11.0582%	27.0864%
High Heating Value, HHV	8.1567%	35.2432%
Carbon	7.7918%	43.0349%
Ash	5.1969%	48.2318%
Avg. Secondary Air Htr. Exit Gas Temp.	5.1884%	53.4201%
Avg. Secondary Air Htr. Inlet Air Temp.	4.7522%	58.1723%
Secondary Air Htr. Exit Flue Gas O2	4.6618%	62.8342%
Oxygen	4.4658%	67.3000%
Avg. Primary Air Htr. Exit Gas Temp.	3.7220%	71.0220%
Auxiliary Equipment Power	2.6509%	73.6729%
Fuel Temp.	2.1789%	75.8517%

Who would have thought that unburned carbon in the fly ash would be the first thing on the list? This is not always the case, nor is it typical, which is why you can imagine this one thing alone caused quite an argument between the parties to the test as to how this was being measured and correction applied. Arguments are common when testing boilers and this particular one was quite contentious; so much so that the test was performed several times and all of the instruments carefully scrutinized.

Moisture in the coal accounts for another 11% of the total uncertainty. This is not surprising and why it is necessary to take as many samples as possible and to send these to a reputable laboratory for analysis.

The heating value, which we might have expected to be the top item on this list, accounts for only 8% of the total uncertainty.

Coal carbon content is almost 7.8%, ash content is 5.2%, and oxygen content is 4.5%. No surprise here.

Three temperature measurements (two in the secondary air and one in the primary air) each contribute about 4% to the total.

O_2 measurement in the stack (from which we infer excess air) contributes another 4%.

The impact of auxiliary power (station service load) in this calculation is a matter of some disagreement. This is discussed briefly in PTC4. Differing parties to a test may argue at length over the details of this contribution, which are beyond the scope of this text. We do include it here and point out that it is only 2.7% of the total uncertainty.

The fuel temperature comes into play with the calculation of sensible heat, which we saw in spreadsheets boiler?.xls.

Gas-Fired Example

We next consider a typical gas-fired boiler test, which can be found on the gas tab in the same spreadsheet as the coal example (uncertainty.xls). This system has different instrumentation on the steam side so that a heat balances around only one feedwater heater is necessary. In this case there are a total of 41 measurements that directly or indirectly impact the calculated boiler efficiency. These include 3 ambient conditions; 13 fuel-related analyses, measurements, or assumptions; 2 air preheater measurements or estimates; 2 flue gas measurements; and 21 steam-related measurements.

The top left section of the spreadsheet is shown in this next figure:

	A	B	C	D	E	F
1	**GAS-FIRED BOILER EFFICIENCY TEST UNCERTAINT**					
2				Measurement Uncer		
3		Test Value		Systematic		
4	TEST UNCERTAINTY (Absolute Basis) (95% Confidence Level)	Mean, $\bar{X}$	Units	B_{Inst} Instrument Systematic Uncertainty	$B_{Spatial}$ Spatial Systematic Uncertainty	$U_{95,SYS}$ Overall Systematic Uncertainty
6	**Ambient Data**					
7	Barometric Pres.	14.638	psia	5.00E-02	0.00E+00	5.00E-02
8	Ambient Relative Humidity	53	%	4.00E+00	0.00E+00	4.00E+00
9	Ambient Dry Bulb Temp.	74	Deg F	5.00E+00	2.00E+00	5.39E+00
10	**Fuel Analysis**					
11	Methane (xCH4)	95.190	% Moles	1.06E-01	0.00E+00	1.06E-01
12	Ethane (xC2)	2.480	% Moles	7.07E-02	0.00E+00	7.07E-02
13	Propane (xC3)	0.440	% Moles	4.95E-02	0.00E+00	4.95E-02
14	Iso-Butane (xIC4)	0.090	% Moles	1.41E-02	0.00E+00	1.41E-02
15	N-Butane (xNC4)	0.090	% Moles	1.41E-02	0.00E+00	1.41E-02
16	Iso-Pentane (xIC5)	0.040	% Moles	1.41E-02	0.00E+00	1.41E-02
17	N-Pentane (xNC5)	0.020	% Moles	1.41E-02	0.00E+00	1.41E-02
18	N-Hexane (xC6)	0.060	% Moles	1.41E-02	0.00E+00	1.41E-02
19	Carbon Dioxide (xCO2)	1.070	% Moles	7.07E-02	0.00E+00	7.07E-02
20	Nitrogen (xN2)	0.510	% Moles	4.95E-02	0.00E+00	4.95E-02

Figure 27. Gas-Fired Uncertainty Spreadsheet (top left)

This is very similar to the coal-fired uncertainty analysis and contains many of the same parameters. Of course, the fuel analysis is different and is provided on a mole (volume) basis rather than a mass basis as with coal. We have the same columns for systematic uncertainty.

G	H	I	J	K	L	M	N	P
Y - REFER TO PTC4 (BOILERS) AND PTC19.1 (UNCERTAINTY)								
tainty Budget				Uncertainty of Test Results				
Random			Total	Corrected Boiler Efficiency			89.07	part
$S_{\bar{X}}$ Standard Deviation of the Mean	$t_{95,\nu}$ Student's t	$U_{95,RAND}$ Random Uncertainty	$U_{95,TOT}$ Total Measurement Uncertainty	θ Absolute Sensitivity	$U_{P1,SYS}$ [kW] Systematic Uncertainty of Corrected Power	$U_{P1,RAND}$ [kW] Random Uncertainty of Corrected Power	U_{P1} [kW] Total Uncertainty of Corrected Power	Contribution to total uncertainty
1.25E-02	1.97E+00	2.46E-02	5.57E-02	2.49E-03	1.24E-04	6.13E-05	1.39E-04	0.1655%
1.00E+00	1.97E+00	1.97E+00	4.46E+00	-6.07E-04	-2.43E-03	-1.20E-03	2.71E-03	0.7311%
1.25E+00	1.97E+00	2.46E+00	5.92E+00	-1.22E-03	-6.57E-03	-3.00E-03	7.22E-03	1.1942%
5.30E-02	1.97E+00	1.04E-01	1.49E-01	-2.50E-02	-2.66E-03	-2.62E-03	3.73E-03	0.8580%
3.54E-02	1.97E+00	6.96E-02	9.93E-02	5.75E-01	4.07E-02	4.00E-02	5.71E-02	3.3570%
2.47E-02	1.97E+00	4.88E-02	6.95E-02	1.15E+00	5.70E-02	5.62E-02	8.00E-02	3.9757%
7.07E-03	1.97E+00	1.39E-02	1.99E-02	1.72E+00	2.43E-02	2.40E-02	3.41E-02	2.5962%
7.07E-03	1.97E+00	1.39E-02	1.99E-02	1.73E+00	2.45E-02	2.41E-02	3.43E-02	2.6033%
7.07E-03	1.97E+00	1.39E-02	1.99E-02	2.30E+00	3.25E-02	3.20E-02	4.56E-02	3.0017%
7.07E-03	1.97E+00	1.39E-02	1.99E-02	2.31E+00	3.26E-02	3.21E-02	4.58E-02	3.0066%
7.07E-03	1.97E+00	1.39E-02	1.99E-02	2.88E+00	4.08E-02	4.02E-02	5.72E-02	3.3614%
3.54E-02	1.97E+00	6.96E-02	9.93E-02	-7.73E-01	-5.47E-02	-5.39E-02	7.67E-02	3.8929%
2.47E-02	1.97E+00	4.88E-02	6.95E-02	-7.72E-01	-3.82E-02	-3.76E-02	5.36E-02	3.2534%

Figure 28. Gas-Fired Uncertainty Spreadsheet (top right)

The summary can be found in the lower right corner:

K	L	M	N	P
AND PTC19.1 (UNCERTAINTY)				
Uncertainty of Test Results				
Corrected Boiler Efficiency			89.07	part
θ Absolute Sensitivity	$U_{P1,SYS}$ [kW] Systematic Uncertainty of Corrected Power	$U_{P1,RAND}$ [kW] Random Uncertainty of Corrected Power	U_{P1} [kW] Total Uncertainty of Corrected Power	Contribution to total uncertainty
-3.16E-02	-1.58E-02	-7.77E-03	1.76E-02	1.8637%
1.69E-03	8.44E-03	4.16E-03	9.41E-03	1.3631%
9.53E-04	1.91E-03	9.39E-04	2.12E-03	0.6477%
1.45E-06	2.18E-05	1.07E-05	2.43E-05	0.0693%
-6.50E-05	-5.44E-02	-2.68E-02	6.07E-02	3.4612%
RSS	0.663	0.310	0.73	100.00%

Figure 29. Gas-Fired Uncertainty Spreadsheet (bottom right)

Here we see that the efficiency is 89.07%±0.73% or that we can say with 95% confidence that it lies between 88.34% and 89.81%. The top contributors to uncertainty are listed in decreasing order in the following table.

Table 55. Uncertainty Contributions

order	measurement	contribution	cumulative
1	Fuel Flow	9.2%	9.2%
2	Assumed Turbine Leakage	8.8%	18.0%
3	Average Air Heater Exit Air Temp.	7.5%	25.5%
4	FW Temp.	6.4%	31.9%
5	Average Air Heater Inlet Gas Temp.	5.0%	36.9%
6	Cold Reheat Temp.	4.7%	41.6%
7	Air Heater Inlet Flue Gas O_2	4.5%	46.1%
8	Propane (in fuel)	4.0%	50.1%
9	Carbon Dioxide (in fuel)	3.9%	54.0%
10	Reheater Spray Flow Rate	3.5%	57.4%
13	Ethane (in fuel)	3.4%	67.5%
14	Nitrogen (xN2)	3.3%	70.8%
28	Methane (xCH4)	0.9%	95.9%

We should first note that each boiler design and test will be different and that the ones shown here are examples. While they are typical, you should not include that methane content in the fuel gas always contributes a mere 0.9% to the total uncertainty or is always #28 on the list, as it is here. It's not surprising that fuel mass flow rate is the #1 contributor to uncertainty in boiler efficiency, accounting for 9.2% of the whole.

The second item on the above list (assumed turbine leakage) here contributes 8.8% of the total uncertainty. That doesn't mean that this *adjustment* impacts the boiler efficiency by 8.8%, only that it impacts the uncertainty in boiler efficiency by 8.8%. The total uncertainty in this particular boiler efficiency test was 0.73%. What this means is that we aren't sure what the rate of leakage from the steam turbine was and the customer did not want to perform a special test to determine it. The parties agreed that it might be as low as half or as large as twice the design amount and so we used this range of values to perform the calculations. This is one of many challenges we run into with performance testing: We should measure that. How much does it cost and how long will it take? That's too much and too long, so do without it.

As with the coal-fired boiler, the steam and flue gas temperatures are also important. So are the steam flows. The impact on uncertainty of fuel gas composition is quite small in this particular test but this is not always the case and should not be assumed from the outset. Look back a line 11 of Figure 27 to see that methane makes up 95.19% of this fuel. It's mostly methane so juggling the other components around doesn't significantly change the overall result.

> *Always try to measure everything you can as accurately as possible and don't presume anything is unimportant until you bring it all together for the final analysis. By that point it's too late to go back and do it all over again because you should have been more careful with one particular measurement.*

Chapter 12. Chemical Reactions

Chemical reactions in any type of combustion can be complex, especially for boilers. Boiler test procedures (e.g., PTC4) are designed in such a way that the details of combustion are reduced to inputs and outputs. We send coal, gas, or oil samples to a laboratory for analysis. We measure simple products of combustion in the stack (CO_2, CO, SO_2, and H_2O). We use the heating values as provided or perhaps calculate these from empirical formulas in the case of coal and oil or tables in the case of gas. The reaction or reactions themselves are rarely considered in this context. "You get what you get," is the approach most often in the world of boiler testing. "Just tell me if it passes and send the bill."

This approach often works... but what if it doesn't? I have tested three boilers that failed and one that passed by a ridiculous margin. Needless to say, heated arguments ensued, either one party demanding payment for damages or the other demanding a massive performance bonus. More than three boilers have failed in my long testing career, but those were due to identifiable defects. These four boilers had no identifiable defects. Several potential defects were suggested but these were dismissed after careful inspection.

In each case the measurements were carefully performed and the instruments were calibrated. Calibration of the instruments was confirmed by on-site procedures (e.g., put all of the temperature probes in a bucket of water to see if they read the same). Something was missing in the analyses for each of these tests in spite of following procedure to the letter. In three of these four cases, chemistry was the problem. In the fourth case (one of the fails), a steam line used for soot blowing had been inadvertently left open and was unaccounted for in the calculations. It was shut off and the test performed again the next day.

For the remaining three cases (two fails and one unreasonable success), the fuel compositions lay outside the normal range (one coal and two fuel oil). Had these been natural gas-fired boilers, we would have adequately captured the chemistry. We *assumed* that the flue gas would contain only N_2, O_2, Ar, CO_2, CO, SO_2, and H_2O but that was not true. We measured O_2, CO_2, CO, and SO_2 and calculated the others making the usual assumptions. To resolve the discrepancies, we had to go back and thoroughly analyze the flue gas, which meant bringing more equipment and subsequent retesting.

In order to analyze chemical reactions, we need a completely different approach than you will find in any PTC plus specialized software. The tool I developed decades ago to perform such analyses (CREST[7]) is described elsewhere[8] and the details are beyond the scope of this text. We will, however,

---

[7] Benton, D. J., "Solution of Complex Thermochemical Equilibria in Symbolic Form," American Society of Mechanical Engineers, Winter Annual Meeting, 1991.
[8] Benton, D. J., *Thermochemical Reactions: Numerical Solutions*, ISBN-9781073417872, Amazon, 2019.

consider a few simple examples to illustrate the importance of such considerations.

This first example is simply the combustion of common coal (50.16% carbon, 43.26% hydrogen, 4.88% oxygen, 0.79% nitrogen, and 0.91% sulfur by mole dry) with standard air at one atmosphere from 500K to 2500K.

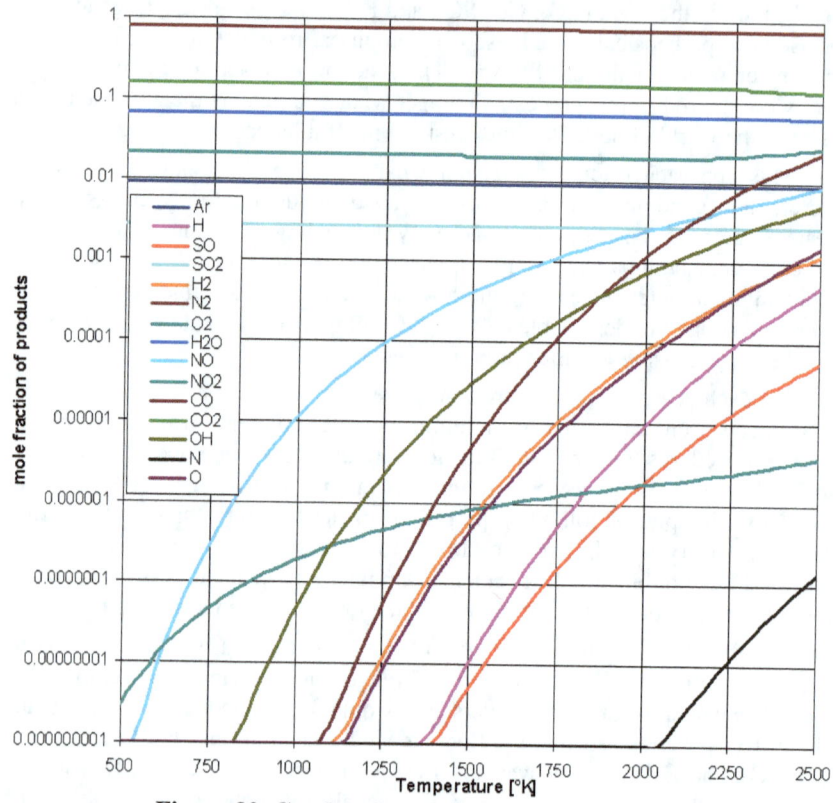

Figure 30. Combustion of Coal at 1 Atmosphere

Combustion temperatures reached 2000K or about 3/4th of the X-axis in this figure. This condition persisted for a significant distance in these boilers due to the design, which was somewhat different from traditional designs. We (and the manufacturer, which was the same for both) were surprised by how this design change, which was made for entirely different reasons, impacted the flue gas composition. Concentrations of O, OH⁻, N, NO, NO_2, and SO persisted downstream of the fireball and into the flue. We didn't measure these gases because they aren't required for the standard analysis and weren't included in the contract documents. When we accounted for these in the flue gas calculations, the results made a lot more sense. Still, both boilers failed the guarantee.

The previous graph shows the impact of combustion temperature on flue gas composition but that is only one variable in this complex problem. We next consider the impact of air/fuel ratio on flue gas composition.

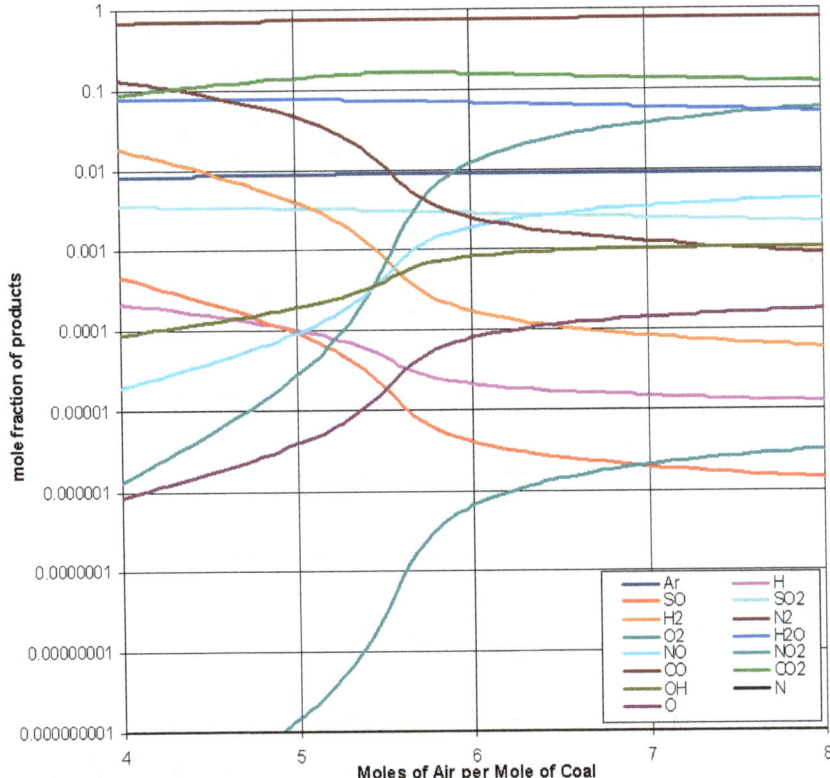

Figure 31. Impact of Air/Fuel Ratio on Flue Gas Composition

The reaction shown in the figure above is also at one atmosphere and the combustion temperature is held constant at 2000K. We see that the composition changes quite rapidly (note that the Y-axis is the log of mole fraction) between 5.5 and 6.0 moles of air per mole of fuel. CO_2 (the green curve) reaches a peak at a molar air/fuel ratio of about 5.7 and falls off beyond this point, while the oxides of nitrogen continue to increase. Neither the manufacturer nor the test team considered these constituents in the original test but did in the subsequent tests.

Gas-Fired Reactions

We next consider the products of combustion for typical natural gas (75.78% methane, 12.50% ethane, 6.25% propane, 3.12% butane, 1.56% pentane, 0.78% hexane). As this next figure shows, the flue gas may contain a variety of gases. In this case we also consider more oxides of nitrogen.

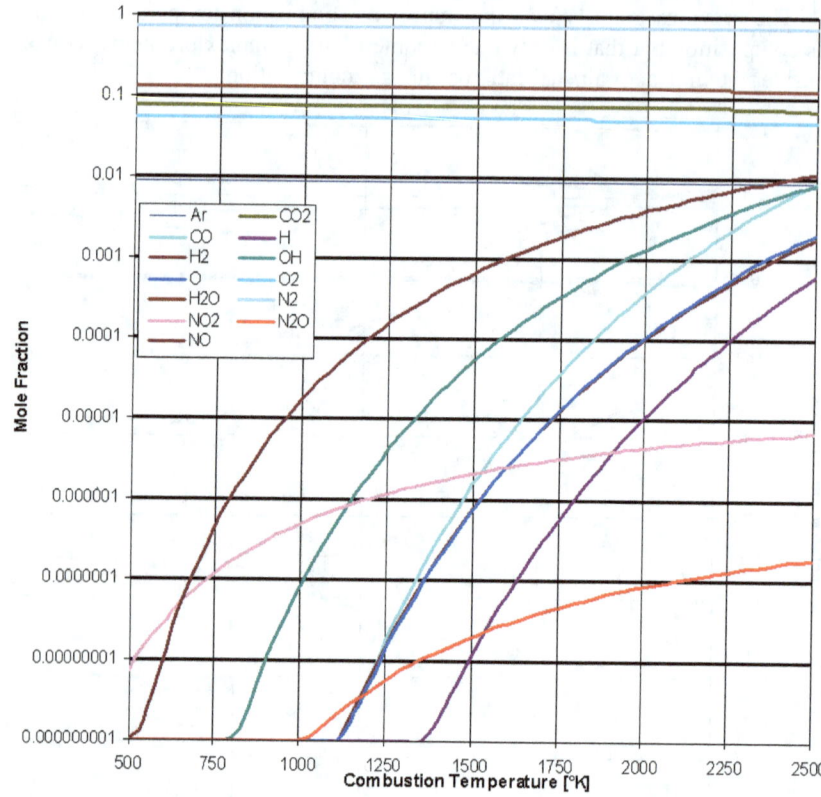

Figure 32. Impact of Temperature on Flue Gas Composition (Natural Gas)

We see in this figure that the flue gas constituents vary considerably (note log axis) and in the case of this third boiler, we must consider this when calculating not only the heat produced by combustion but also that carried away into the atmosphere by the flue gas. The details of these last three examples can be found in reactions.xls

Appendix A. Higher vs. Lower Heating Value

There is much confusion and disinformation surrounding lower vs. higher heating value of fuels and also net vs. gross calorific value. The distinction is whether or not the resulting water is in the liquid or vaporous state. Combustion of hydrogen (or any hydrocarbon) produces water (H_2O). If this is in the *vaporous* state, then we obtain the *lower* heating value. If this is in the *liquid* state, then we obtain the *higher* heating value.

While some bomb calorimeters may end up with water in the liquid state after sufficient cooling, no actual combustion process (such as a boiler or gas turbine) operating on Earth will have liquid water in the exhaust, which will be far above the boiling point. Perhaps if you were to operate such a device in Antarctica during a blizzard, you might have some liquid in the exhaust.

There is some history here... It was thought by some (unfamiliar with chemistry and thermodynamics) that boiler manufacturers who used lower heating value were exaggerating the efficiency of their equipment, while those using higher heating value were not. This is a misunderstanding, not science.

Also, some bomb calorimeters do end up with water in the liquid state and, therefore, inherently yield a higher heating value, which must be corrected downward to account for water in the vapor state. Some consumers feared they were being hoodwinked by calorimetry reporting higher or lower heating values, which might have been true in some cases, but has nothing to do with chemistry or science. While some official documents (for example ASME and DoE plus many Canadian and European standards) might insist on using higher heating value, engineers should follow the science whenever possible.

Figure 33. It's All About Condensation

NOTE: If you are using the NASA Glenn gas properties (Appendix D)—as is the case in all the examples contained herein—water vapor is in the gaseous state, not the liquid state, and so you must use the lower heating value of the fuel in your calculations unless you specifically account for the latent heat of vaporization; otherwise, you will be introducing a property reference error.

Appendix B. Coal Heating Values

Dmitri Ivanovich Mendeleev (1834-1907) Russian chemist and inventor, formulated the Periodic Law, and created an early version of the periodic table of elements. He was financially motivated to accurately quantify the value of coal from various sources. In other words, he developed the science to assure people were getting their money's worth and not being cheated! There has is much money to be made and lost in buying and selling coal; so much effort has been invested in evaluating it or determining, "How much heat will it produce?"

There are several laboratory techniques for analyzing coal—far more than mere *bomb calorimetry* (burn a sample in a heavy iron container and measure the heat generated, often using an insulated water bath to facilitate temperature measurement. While this is perhaps the most straightforward measure, it isn't necessarily the most accurate. Any measurement technique will have some uncertainty and challenges. Don't presume that a heating value measured with a bomb calorimeter will necessarily be more accurate than one calculated from a chemical breakdown analysis.

Table 56. Typical Coals

Constituent	PRB West Seam	PRB East Seam	Lee Ranch	Black Thunder	Antelope Mine	North Antelope	Jacobs Ranch
Carbon	51.47%	49.84%	52.37%	51.04%	51.35%	51.47%	49.52%
Hydrogen	3.43%	3.34%	3.92%	3.49%	3.59%	3.43%	3.57%
Nitrogen	0.66%	0.71%	0.85%	0.58%	0.78%	0.66%	0.63%
Oxygen	12.74%	12.28%	8.68%	11.97%	12.08%	12.71%	13.07%
Sulfur	0.20%	0.22%	0.95%	0.32%	0.24%	0.20%	0.43%
Chlorine	0.01%	0.01%	0.02%	0.01%	0.01%	0.00%	0.01%
Ash	4.40%	4.60%	18.51%	5.30%	5.25%	4.44%	5.47%
Moisture	27.10%	29.00%	14.70%	27.30%	26.70%	27.10%	27.31%
Total	100.01%	100.00%	100.00%	100.01%	100.00%	100.01%	100.01%
Calculation	BTU/lbm	BTU/lbm	BTU/lbm	BTU/lbm	BTU/lbm	BTU/lbm	BTU/lbm
HHV	8870	8555	9200	8800	8800	8870	8773
LHV	8273	7947	8688	8196	8193	8273	8161
Dulong	8634	8377	9414	8673	8768	8636	8421
Boie	8916	8648	9529	8921	9014	8917	8685
Grummel & Davis	8522	8255	9309	8547	8641	8524	8311
Mott & Spooner	8745	8484	9458	8772	8868	8747	8537
Mason & Ghandi	7385	7017	8563	7388	7502	7387	7153
Mendeleev	8106	8482	8341	8189	8127	8199	7220
maximum	8916	8648	9529	8921	9014	8917	8773
average	8431	8221	9063	8436	8489	8444	8158
minimum	7385	7017	8341	7388	7502	7387	7153
standard deviation	507	532	461	502	508	499	630
C/H	15.01	14.92	13.36	14.62	14.30	15.01	13.87

A variety of coals and calculations can be found in spreadsheet coal.xls in the online archive in the examples folder and listed in the preceding table. Articles abound on the various calculations and can be found. A good overview has been provided by Mason and Gandhi.[9] Eight different values are listed in the table. Two of these were reported by laboratory measurement and six are calculated. All of the formulas are embedded in the spreadsheet. The maximum, minimum, average, and standard deviation are shown at the bottom of the table and are also typical. This illustrates the unavoidable uncertainty in the calculations, as discussed in Chapter 4. A simple formula for HHV and LHV in BTU/lbm provided in this reference is the following (for kJ/kg multiply by 2.326):

$$HHV = 15725 x_C + 47758 x_H + 5048(x_S - x_O) - 1165 x_{H_2O} \quad (B.1)$$

$$LHV = HHV - 9114 x_H - 1032 x_{H_2O} - 39 x_O \quad (B.2)$$

---

[9] Mason, D. M. and K. Gandhi, "Formulas for Calculating the Heating Value of Coal and Coal Char: Development, Tests and Uses," *Environmental Science*, 1980.

Appendix C. Natural Gas Heating Values

The heating value of natural gas is calculated from the composition. Rarely is this determined through actual combustion in a laboratory, as this can be quite problematic. The reference quantities are well-established and can be calculated from the information provided in the NASA Glenn Report (see appendix D). Some common values are listed below. The 46 most common constituents are listed in spreadsheet natural_gas.xls, which may be found in the examples folder.

Table 57. Heating Values for Some Natural Gas Constituents

name	formula	MW	BTU/lb-mole		kJ/kg-mole	
			LHV	HHV	LHV	HHV
Hydrogen	H2	2.016	273.8	324.2	636.9	754.1
Methane	CH4	16.043	909.4	1010.0	2115.3	2349.3
Ethyne	C2H2	26.038	1256.1	1300.0	2921.7	3023.8
Ethene (Ethylene)	C2H4	28.054	1412.1	1323.2	3284.5	3077.8
Ethane	C2H6	30.070	1618.7	1769.6	3765.1	4116.1
Propene (Propylene)	C3H6	42.081	2059.4	1926.1	4790.2	4480.1
n-Propane	C3H8	44.097	2314.9	2516.1	5384.5	5852.4
Butenes	C4H8	56.108	2543.6	2742.9	5916.4	6380.0
i-Butane	C4H10	58.123	3000.4	3251.8	6978.9	7563.9
n-Butane	C4H10	58.123	3010.8	3262.3	7003.1	7588.1
Butanes	C4H10	58.123	3010.8	3262.3	7003.1	7588.1
1-Pentene	C5H10	70.134	3158.2	3407.4	7346.0	7925.6
i-Pentane	C5H12	72.150	3699.0	4000.9	8603.9	9306.1
n-Pentane	C5H12	72.150	3706.9	4008.9	8622.2	9324.7
Neopentane	C5H12	72.150	3706.9	4008.9	8622.2	9324.7
Pentanes	C5H12	72.150	3706.9	4008.9	8622.2	9324.7
Benzene	C6H6	78.114	3137.8	3270.0	7298.5	7606.0
Cyclohexane	C6H12	84.161	3656.8	3955.8	8505.7	9201.2
n-Hexane	C6H14	86.177	3856.6	4205.4	8970.5	9781.8
n-Heptane	C7H16	100.204	4465.7	4864.3	10387.2	11314.4
n-Octane	C8H18	114.231	5074.9	5523.4	11804.2	12847.4
n-Nonane	C9H20	128.258	5683.2	6181.6	13219.1	14378.4
n-Decane	C10H22	142.285	6294.7	6842.8	14641.5	15916.6
Carbon Monoxide	CO	28.010	320.5	320.5	745.5	745.5
Carbon Dioxide	CO2	44.010	0.0	0.0	0.0	0.0
Sulfur Dioxide	SO2	64.065	0.0	0.0	0.0	0.0
Hydrogen Sulfide	H2S	34.082	586.8	637.1	1364.9	1481.9

Appendix D. Flue Gas Properties

The prime source for gas properties is the NASA Glenn report.[10] Most other references (e.g., ASME PTC22) are derived from these or the JANAF Report[11], which was the predecessor. Note that specific heat and enthalpy are mass-weighted, not mole-weighted. Common specific heats are shown in the following figure:

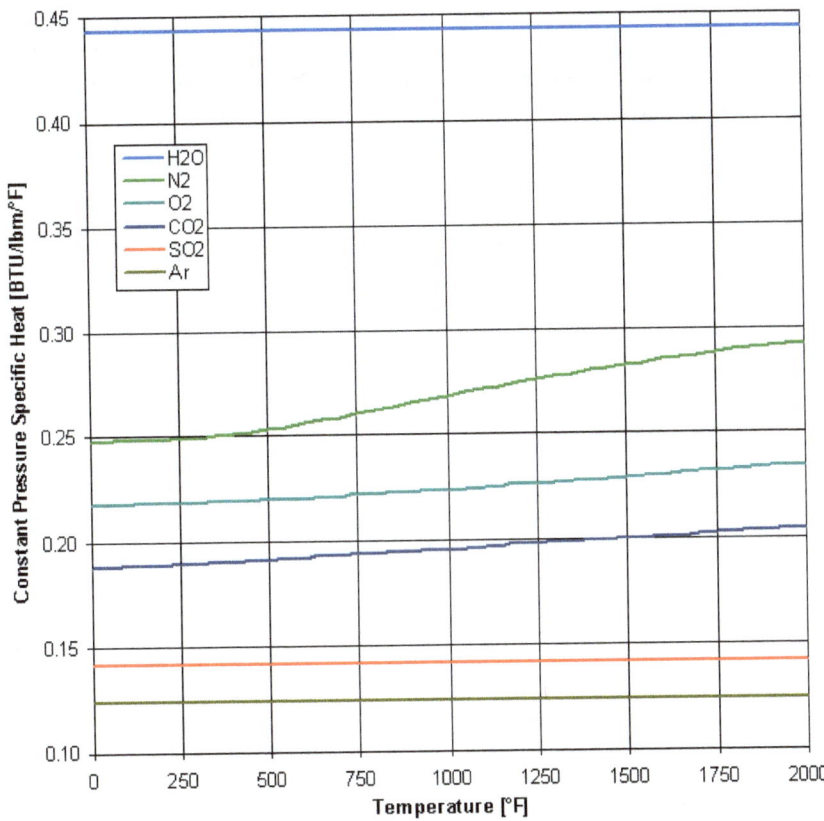

VBA code for these properties can be found in spreadsheet boiler1.xls, which is included in the online archive. Complete details and additional spreadsheets can be found in the archive accompanying my book, *Thermodynamic and Transport Properties of Fluids*.

---

[10] McBride, B. J., Zehe, M. J., Gordon, S., "NASA Glenn Coefficients for Calculating Thermodynamic Properties of Individual Species," NASA Report No. 211556, 2002.

[11] Chase M. W., Davies, C. A., Downey, J. R., Frurip, D. J., McDonald, R. A., and Syverud, A. N., "JANAF [Joint Army, Navy, Air Force] Thermochemical Tables, Third Edition,", Journal of Physical Chemistry Reference Data, Vol. 14, Supplement 1, 1985.

Enthalpy of the flue gas constituents (which must be on a mass, not mole basis), is obtained by integrating the (constant pressure) specific heat, C_P, with respect to temperature:

$$h = \int_{Tref}^{T} C_p dT \qquad (D.1)$$

This has been performed analytically for the constituents included in spreadsheet boiler1.xls for computational efficiency. A free Excel Add-In providing convenient access to all of the gases and properties in the NASA Glenn Report can be found at the link beneath the Preface. For the range 0°F to 4500°F the following are adequate:

```
hN2=(9.33684479355E-6*T+0.250024091127)*T
hO2=(8.63368146203E-6*T+0.230677461577)*T
hAr=0.124278760622*T
hCO2=((-3.85945528886E-9*T+0.000037149663185)*T+0.211176364782)*T
hSO2=((-2.23936059887E-9*T+2.02804245109E-5)*T+0.156596173768)*T
hH2O=(3.93014501942E-5*T+0.440265936316)*T
```

Appendix E. Moist Air Properties

The only properties of moist air considered are those developed by Hyland & Wexler[12][13][14], and refined by Nelson & Sauer.[15] The more recent formulation developed by Hermann, Kretzschmar, and Gatley[16][17] do not constitute a substantive improvement, merely an academic one; so there is little point implementing these. Moist air properties appear in various editions of the *ASHRAE Handbook of Fundamentals*. Beware that the equations in many editions of this otherwise excellent reference are wrong in that the equations contained therein don't produce the tabulated results.

In 1984 this author was part of a Cooling Technology Institute (CTI) task force investigating discrepancies in the published properties of moist air. The National Bureau of Standards (NBS)—now the National Institute of Standards and Technology (NIST)—lost Hyland & Wexler's original reports; however, a copy still existed in the Library of Congress (LoC). A colleague, Al Feltzin, went to the LoC and made a photocopy of the original reports. The tabulated values in the ASHRAE handbook are correct, but not all of the equations are, especially before 1993.

Formulations consistent with Hyland & Wexler, along with code plus an Excel® Add-In can be found in my book, *Evaporative Cooling*, and on my web site listed in the foreword. The code, calculations, and spreadsheets can be found in the online archive in folder examples\boiler1.xls. The humidity ratio, W, is the ratio of the mass of water to mass of dry air and is given by:

$$W = \left(\frac{MW_{H2O}}{MW_{AIR}}\right)\left(\frac{fP_{SAT}}{P_{BARO} - fP_{SAT}}\right) \qquad (E.1)$$

Where MW_{H2O} and MW_{AIR} are the molecular weights of water and air, respectively. P_{SAT} and P_{BARO} are the saturation and barometric pressures,

---

[12] Hyland, R. W., Wexler, A., and Stewart, R., "Thermodynamic Properties of Dry Air, Moist Air and Water and SI Psychrometric Charts," ASHRAE RP-216 and RP-25, 1983.

[13] Hyland, R. W. and Wexler, A., "Formulations for the Thermodynamic Properties of the Saturated Phases of H2O from 173.15 K to 473.15 K," ASHRAE Trans., Vol. 89, pp. 500-519, 1983.

[14] Hyland, R. W. and Wexler, A., "Formulations for the Thermodynamic Properties of Dry Air from 173.15 K to 473.15 K, and of Saturated Moist Air from 173.15 K to 372.15 K, at Pressures to 5 MPa," ASHRAE Trans., Vol. 89, pp. 520-535, 1983.

[15] Nelson, H. F. and Sauer, H. J., "Formulation of High-Temperature Properties for Moist Air," HVAC&R Research Vol. 8, pp. 311-334, 2002.

[16] Herrmann, S., Kretzschmar, H.-J., and Gatley, D. P., "Thermodynamic Properties of Real Moist Air, Dry Air, Steam, Water, and Ice," HVAC&R Research, 2009.

[17] Herrmann, S., Kretzschmar, H.-J., and Gatley, D. P., "Thermodynamic Properties of Real Moist Air, Dry Air, Steam, Water, and Ice - Final Report," ASHRAE RP-1485, 2009.

respectively. The enhancement factor, f, is the ratio of the effective partial pressure of water vapor in air at the saturation point to the saturation pressure of steam alone (no air present). The enhancement factor varies with temperature and barometric pressure and is shown in the figure below for one atmosphere:

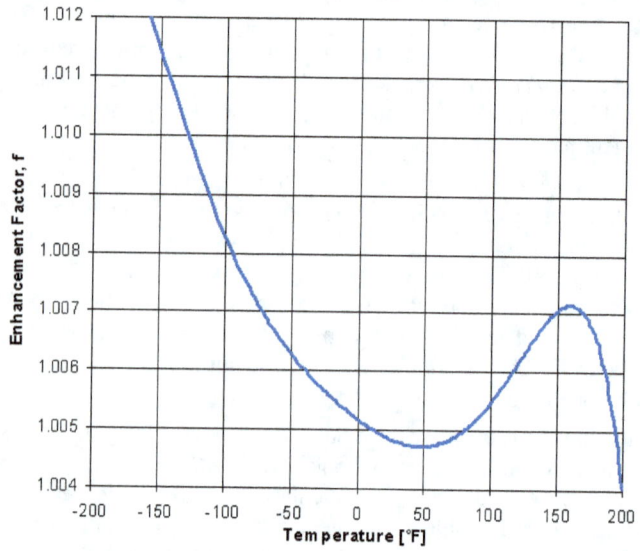

The saturation pressure for water vapor (against water liquid without air present) is shown in this next figure:

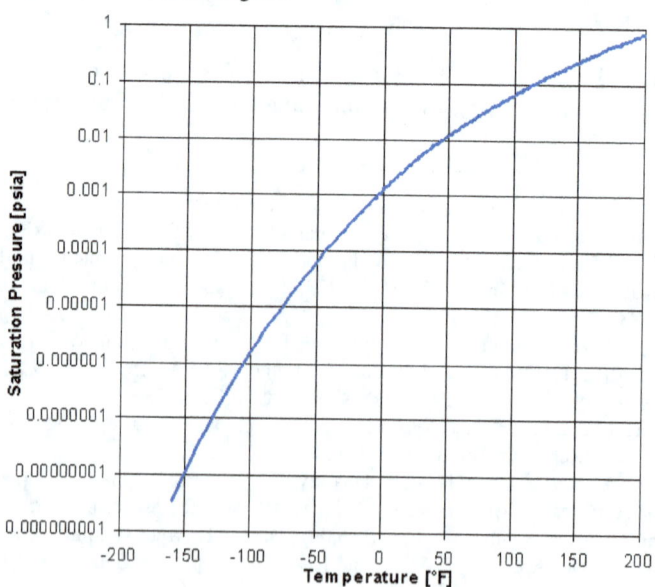

The resulting humidity ratio is then:

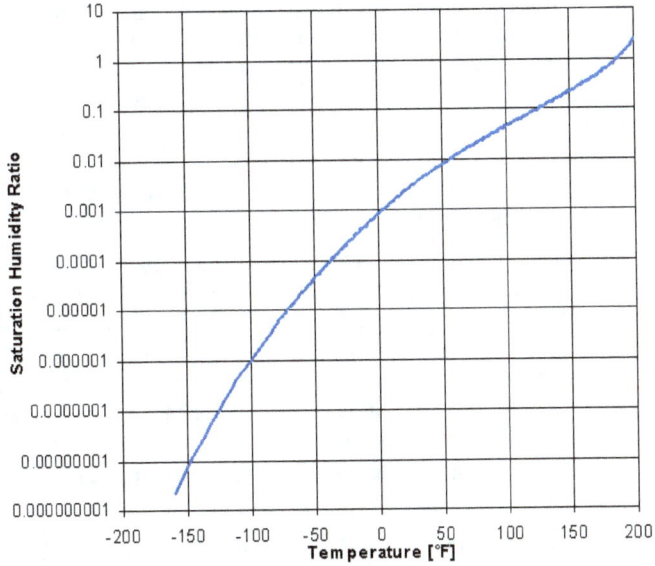

The enthalpy of moist air (per pound of dry air) is given by:

$$h = h_A + Wh_G = 0.24\,T + W(1061 + 0.444\,T) \qquad (E.2)$$

where h_A and h_G are the enthalpies of dry air and water vapor, respectively. The enthalpy of water vapor is given by:

$$h_G = 1061 + 0.444\,T \qquad (E.3)$$

Note that this is the enthalpy of saturated water vapor, not the latent heat (h_{FG}) as is sometimes thought and even cited in the literature.

Note that you will need rigorous psychrometric calculations for ambient air, but DO NOT use these in the flue, as they are meaningless above the boiling point (212°F/100°C). In that case, use NASA Glenn, but be aware that the references are different. ASHRAE (Hyland & Wexler) reference liquid water at the triple point and NASA Glenn reference vapor water at standard conditions. If you do not consider this difference in the energy balance, you will get erroneous results.

Appendix F. Steam Properties

The first seamless steam property formulation was developed by Keenan, Keyes, Hill, and Moore[18] in 1969. This excellent work was rarely used at the time and has not been used in decades. The formulation of Meyer, McClintock, and Silvestri, published in 1967[19] and endorsed by the ASME has been much more widely used. A significant improvement was made by Haar, Gallagher, and Kell[20] in 1984. The 1984 formulation was very similar to the 1969, in that it was mathematically elegant and continuous, but it too was rarely used, as many computers came without a floating-point processor at that time.

Three formulations have been published by the International Association for the Properties of Water and Steam.[21] These are called: IF-67[22], SF-95[23], and IF-97[24]. The designations IF and SF indicate Industrial and Scientific Formulations, respectively. The industrial formulations are in terms of pressure and temperature, while the scientific formulation is in terms of temperature and density. While the industrial formulations are less accurate and mathematically sloppy, they have been embraced throughout the industry. The choice of formulation is most often driven by the steam turbine manufacturer and not by the boiler manufacturer.

This author has developed the most up-to-date and extensive formulation of steam properties, which are not only continuous but also extend to much higher temperatures and pressures.[25] A free Excel Add-In for non-commercial use is available at the link below the Preface that includes all of these formulations. You will also find a VBA implementation of the 1967 properties inside spreadsheet boiler3.xls.

---

[18] Keenan, J. H., Keyes, F. G., Hill, P. G., and Moore, J. G., *Steam Tables*, John Wiley & Sons, Inc., 1969.

[19] Meyer, C. A., McClintock, R. B., Silvestri, G. J., and Spencer, R. C., Jr., *Thermodynamic and Transport Properties of Steam*, American Society of Mechanical Engineers, 1967.

[20] Haar, L., Gallagher, J. S., and Kell, G. S., *Steam Tables*, NBS/NRC printed by Hemisphere, distributed by McGraw-Hill, 1984.

[21] This is a link to their web site http://www.iapws.org/

[22] Meyer, C. A., McClintock, R. B., Silvestri, G. J., and Spencer, R. C., Jr., *Thermodynamic and Transport Properties of Steam*, American Society of Mechanical Engineers, 1967.

[23] Wagner, W., and Pruß, A., "The IAPWS Formulation 1995 for the Thermodynamic Properties of Ordinary Water Substance for General and Scientific Use," Journal of Physical Chemistry, Ref. Data 31, pp. 387-535, 2002.

[24] Research and Technology Committee on Water and Steam in Thermal Power Systems, *ASME Steam Properties for Industrial Use*, The American Society of Mechanical Engineers.

[25] Benton, D. J., *Steam 2020: to 150 GPa and 6000 K*, Amazon, 2020.

Appendix G. Brent's Method

The desired solution to an equation is not always a root or zero point. We often seek a minimum or maximum. Some methods only work for functions that cross over the y-axis (i.e., have both positive and negative values). Brent's method searches for a minimum.[26] Brent's method arises from a quadratic approximation or interpolant (i.e., the second order Lagrange interpolating polynomial). The iteration begins with three points (x_1, x_2, x_3), presumably bracketing the minimum. The function is evaluated at these three points (y_1, y_2, y_3). The resulting polynomial is:

$$y = \frac{(x-x_2)(x-x_3)y_1}{(x_1-x_2)(x_1-x_3)} + \frac{(x-x_1)(x-x_3)y_2}{(x_2-x_1)(x_2-x_3)} + \frac{(x-x_1)(x-x_2)y_3}{(x_3-x_1)(x_3-x_2)} \quad (G.1)$$

Take the derivative of Equation G.1 with respect to x, set this to zero, and solve to obtain:

$$x = \frac{1}{2} \frac{(x_2^2 - x_3^2)y_1 + (x_3^2 - x_1^2)y_2 + (x_1^2 - x_2^2)y_3}{y_1(x_2 - x_3) + y_2(x_3 - x_1) + y_3(x_1 - x_2)} \quad (G.2)$$

If the revised estimate is to the left x_2, swap out x_3; otherwise, swap out x_1. Continue until the improvement or the step size is less than some tolerance. A typical problem is illustrated in the following figure:

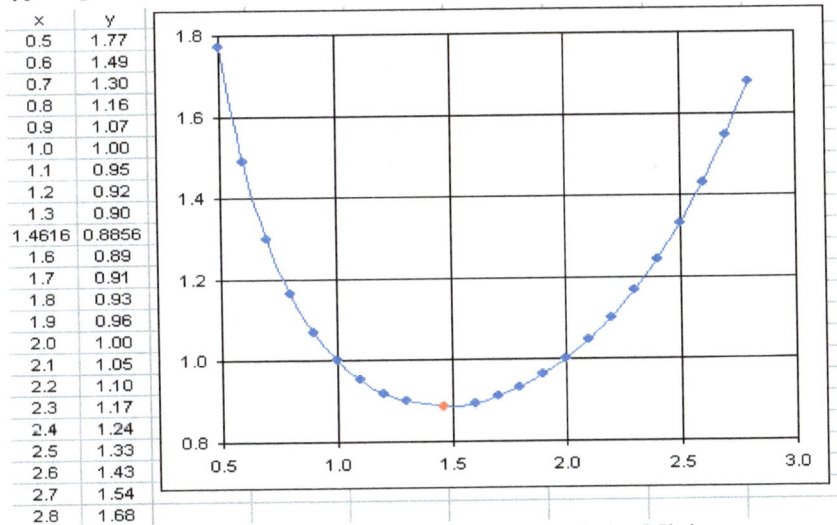

Figure 34. Illustration of a Function Exhibiting a Minimum

---

[26] Brent, R. P., *Algorithms for Minimization without Derivatives*, Chapter 4: An Algorithm with Guaranteed Convergence for Finding a Zero of a Function, Prentice-Hall, Englewood Cliffs, NJ, 1973.

The algorithm is simple to implement:
```
Function Brent(x1 As Double, x3 As Double) As Double
  If (x3 < x1) Then
     a = x3
  Else
     a = x1
  End If
  If (x1 > x3) Then
     b = x1
  Else
     b = x3
  End If
  x1 = a
  x3 = b
  x2 = (x1 + x3) / 2
  y1 = Combustion(x1)
  y2 = Combustion(x2)
  y3 = Combustion(x3)
  If (y1 < y3 And y1 < y2) Then
    Brent = x1
    Exit Function
  End If
  If (y3 < y1 And y3 < y2) Then
    Brent = x3
    Exit Function
  End If
  For iter = 1 To 99
    x = y1 * (x2 - x3) + y2 * (x3 - x1) + (x1 - x2) * y3
    If (Abs(x) < 0.000000001 * (x3 - x2)) Then Exit For
    x = ((x2 * (x2) - x3 * x3) * y1 + (x3 * x3 - x1 * x1) * y2 + (x1 * x1 - x2 * (x2)) * y3) / x / 2
    If (x < a) Then x = a
    If (x > b) Then x = b
    If (x < x2 - 0.000000001) Then
       x3 = x2
       y3 = y2
    ElseIf (x > x2 + 0.000000001) Then
       x1 = x2
       y1 = y2
    Else
       Exit For
    End If
    x2 = x
    y2 = Combustion(x2)
  Next iter
  Brent = x2
End Function
```
This example converges in 5 iterations requiring only 8 function calls

Appendix H. Units

The practicing engineer must be comfortable (and competent) using whatever system of units are expected by the customers who are paying the bills. More often than not with boilers, these will be U.S. Customary (i.e., English) units, rather than SI. I have worked on many projects where all of the documents were in SI units except for the boiler or the cooling tower, which were in English units. There is already more than enough tension and conflict when building, testing, operating, or fixing a power plant without you adding your fussy unit preference and silly reasons, which nobody wants to hear. My advice is to work with whatever units are in the contract documents and avoid the arguments, which are pointless. Below is an excerpt from the only numerical example in the guiding document, which uses English units:

ASME PTC 4-2008
Table B-4-1   Output

	Parameter	Flow, W, Klbm/hr	Temperature, T, °F	Pressure, P, psig	Enthalpy, H, Btu/lbm	Absorption, Q, MKBtu/hr W × (H−H1)/1,000
1	Feedwater (Excluding SH Spray)	433.774	439.5	1,676.6	419.44	
2	SH Spray Water: Ent 1 to Calc HB	0	312.5	2,006.4	284.26	3.573
3	Ent SH-1 Attemp	433.774	0.0	330.7	0.00	
4	Lvg SH-1 Attemp	460.205	0.0	330.7	0.00	
5	SH-1 Spray Water Flow	26.431	W3 × (H3−H4) / (H4−H2) or W4 × (H3−H4) / (H3−H2)			
6	Ent SH-2 Attemp	460.205	0.0	707.9	0.00	
7	Lvg SH-2 Attemp	460.205	0.0	707.9	0.00	
8	SH-2 Spray Water Flow	0.000	W6 × (H6−H7) / (H7−H2) or W7 × (H6−H7) / (H6−H2)			

Figure 35. Excerpt from Appendix B of ASME PTC4-2008

If I can refrain from fussing about Btu vs. BTU (which is an acronym and should be all caps), you can refrain from fussing about Kelvins vs. °K.

Appendix I: Modified Broyden's Method
Applications of a Hybrid Derivative-Free Algorithm for Locating Extrema

I presented this paper at the Society of Industrial and Applied Mathematics, Southeastern Regional Seminar, held in Cullowhee, North Carolina, April 12-13, 1991. Computers have come a long way since then but the algorithms described herein are still the best I have found.

Abstract

Applications of a hybrid derivative-free algorithm for locating extrema of nonlinear functions of several variables based on Broyden's method is presented in which the problems of starting values and extraneous entrapment are addressed. The principal intended application of the algorithm is to find solutions to simultaneous nonlinear equations. The main objective of the algorithm is to minimize the number of function evaluations for problems where the equations are computationally intensive or partial derivatives cannot be determined analytically. Four examples drawn from diverse fields are given for illustration. Comparisons are made to the Newton-Raphson, conjugate-gradient, and steepest-descent methods.

Nomenclature

A=rectangular matrix having M columns and N rows
B=column matrix having M elements
F=a function of several variables
M=the number of residuals ($M \geq N$)
N=the number of unknowns
R=residual column matrix having M elements
X=unknown column matrix having N elements
superscript
T=matrix transpose
subscripts
N=new or current value
O=old or previous value

Introduction

Many practical problems can be cast into the form of a search for extrema of a function of several variables. A common function is the sum of squared residuals, in which case the extrema of interest are the roots of simultaneous equations. Methods abound which require knowledge of the partial derivatives. Many of these derivative-based methods can be adapted by using finite differences to solve problems where the partial derivatives cannot be analytically determined. Such implementations are impractical when the function is computationally intensive.

Derivative-based methods such as the Newton-Raphson discard at each step all information previously learned about the behavior of the function except the current location. Even the Conjugate-Gradient method when applied to nonlinear problems may only preserve one previous direction of search. When the function evaluation is computationally intensive it is essential that as much information as possible about the behavior of the function learned from previous evaluations be preserved and utilized.

Broyden's method is very attractive when considered from this perspective. It does not require knowledge of the partial derivatives, nor does it attempt to compute them directly. Furthermore, Broyden's method preserves all of the information learned about the behavior of the function for the last $N+1$ steps where N is the number of unknowns.

Four enhancements to Broyden's method were made to arrive at the present algorithm: a method for selecting starting values, step length control, hybrid search algorithm, and a method for escaping from extraneous entrapment.

The Basic Method

Given a set of N unknowns represented by the column matrix X and a corresponding set of M residuals represented by the column matrix R, the least-squares function would be $F=R^T R$. The extrema of F occur at the locations where $\mathit{MF}/\mathit{MX}=0$. If the residuals, R, were linear functions of the unknowns, X, then the function, F, would be quadratic and its contours would plot as ellipsoids. This linear case could be described by Equations I.1 and I.2:

$$R = A^T X + B \qquad (I.1)$$

$$F = R^T R = X^T A A^T X + 2 B^T A^T X + B^T B \qquad (I.2)$$

where A is a rectangular matrix having M columns and N rows and B is a column matrix having M elements.

Broyden (1969) reasoned that A and B should be selected such that exact agreement would be preserved for the previous $N+1$ steps. Assuming that no two of the previous $N+1$ Xs are the same, there should be a unique solution to the resulting $M(N+1)$ equations for the elements of A and B. Ignoring for the moment how this sequence of Xs might be obtained, the matrices A and B can be sequentially updated using the following algorithm:

$$A_N = \frac{A_O + [(R_N - R_O) - A_O(X_N - X_O)](X_N - X_O)^T}{(X_N - X_O)^T (X_N - X_O)} \qquad (I.3)$$

$$B_N = R_N - A_N^T X_N \qquad (I.4)$$

where the subscripts N and O refer to *new* and *old* respectively—or the current step and the previous one. Equations I.3 and I.4 can be verified by substitution into Equation I.1 with the *new* and *old* subscripts added. If A and B are initialized to zero and $N+1$ unique starting values of X are selected, then after $N+1$ function evaluations and updates, matrices A and B will be uniquely defined and the search for a solution could proceed.

The interesting property of Equation I.3, which led Broyden to this selection is that, the change in A is only in the direction of the last step in X. That is, the only information about the behavior of the function, which is added to A at each step, is its variation along the current search direction. All of the information about the function in the $N-1$ directions orthogonal to the current search direction remains intact; thus, it is a *rank-one* update method.

Broyden used this algorithm for obtaining and updating matrices A and B, along with Newton's method to search for the extrema. Thus in its original form, Broyden's is a quasi-Newton method (Morè and Sorensen (1984) discuss Newton and quasi-Newton methods in some detail.). The following calculus can be applied to the matrices in order to illustrate this procedure.

$$\frac{\partial R}{\partial X} = A^T$$
(I.5)

$$R_O = A_O^T (X_O - X_N)$$
(I.6)

Matrix A contains the partial derivatives of the residuals, R, with respect to the unknowns, X. Thus matrix A is the *Jacobian* of R with respect to X.

As indicated by Equation I.6, the gradient of the function lies along the direction AR; therefore, most any gradient search method could be implemented and updated using Broyden's method for determining the Jacobian. Ortega and Rheinboldt (1970) discuss on an analytical level a number of methods which could be applied at this point. Actual selection of a practical method, which will produce satisfactory stable results for a wide range of problems, is quite another matter.

The Modified Method

Nonlinear simultaneous equations may have no solution, one solution, or many solutions. The most helpful physical analogy is that of a relief map of the Earth's surface where the unknowns are latitude and longitude and the function is the elevation with respect to mean sea level. No conceivable practical method could hope to locate Mt. Everest or the Marianas Trench regardless of the starting values. While it is reasonable to search for local extrema, it is fortuitous to locate the global extremum—assuming one does exist. Given this analogy it is understandable that no practical algorithm can be expected to locate even a local extremum in every case. Fletcher (1987) discusses these and other problems related to locating extrema in more detail.

Selection of Starting Values

This geographical analogy illustrates the necessity of limiting the region to be searched for extrema. In the present algorithm, a minimum and maximum value for each element in X must be supplied. This not only provides an extent to the range of X, but it also serves as an indication of the scale. Any change in X, which is on the order of the machine precision when compared to the range of X, is considered negligible. One logical choice for the $N+1$ starting values of X would be the center plus the N evenly distributed surrounding values inside the hypercube defined by the specified range of X.

If the function at the central point is greater than at the surrounding points, then the first iteration would direct the search outside of the range of X. If this occurs the range is bisected such that the new center point is mid way between the previous center and the surrounding point corresponding to the least value of the function. If this bisection is unsuccessful after sufficient attempts so as to diminish the subrange of any element of X to the previously determined negligible level, the search is abandoned.

Step Length Control

The unmodified method often results in unstable iterations. Not only is it necessary to confine X to the specified range, it is also necessary to damp the iteration or, as in this case, apply a step length control algorithm. Ortega and Rheinboldt discuss several step length algorithms. The parabola method defined by the current location, one *close* point, and the next step prescribed by the unmodified method has proven to be as successful as any tested. Using the unmodified point as an outer limit on the step length arises from the observation that the unmodified method has a strong tendency to overshoot.

Hybrid Search Algorithm

Because Broyden's is basically a quasi-Newton method, the search proceeds in much the same direction as with *Newton-Raphson* (*NR*). In cases where the *NR* method would fail to locate an extremum, most likely Broyden would also. Broyden's method can also be viewed as a means by which to obtain the Jacobian (matrix A). If the Newton iteration is not successful, the Jacobian can be used to implement other methods. The method of *steepest descent* (*SD*) is more robust, but converges less rapidly than *NR*. When the *NR* iteration fails to result in a reduction of the residual, the direction of steepest descent is searched.

In the present algorithm, the *Conjugate-Gradient* (*CG*) method with the restart procedure recommended by Powell (1977) is also used to supplement the *NR/SD* iteration. The only information added to the Jacobian by Broyden's update is along the search direction. Information about the character of nonlinear functions in directions orthogonal to the search direction can be essential to locating extrema. The *CG* method provides a systematic procedure for searching other directions. In the present algorithm, the *NR*, *SD*, and *CG* methods are used alternately as each ceases to provide continual reduction of the residual.

Escape from Extraneous Entrapment

If N directions have been searched without further improvement, then either a local extremum has been found or extraneous entrapment has occurred. Whether the current location is a local extremum or a nuisance of finite-precision arithmetic can be partially determined by examining the history of matrix A. For nonlinear problems the character of A can change substantially as the search proceeds.

The unmodified Broyden update to A replaces the information along the direction of the current step—thus discarding the previous information along this direction. If an *old* copy of A is retained along with the *new* copy and the search direction indicated by the old A is away from that indicated by the new A (viz. the dot product of the column matrices is less than or equal to zero), then the iteration may have skipped over an inflection point. In this case a search is conducted along the direction connecting these two provisional new values of X.

Extraneous entrapment can sometimes be corrected by arbitrarily perturbing the solution away from the current location to see if it will return to the same point. After this perturbation has been attempted without success in N directions the procedure is abandoned.

Extension to Least Squares

In the case where $M>N$, Equation I.7 must be pre-multiplied by A. The simultaneous nonlinear equations are then solved in the least-squares sense. For most problems the stability of the method also improves when this multiplication is performed even in the case of $M=N$. Therefore, in the present algorithm it is done regardless of the values of N and M.

Comparison to Other Methods

The present derivative-free *enhanced* Broyden (*EB*) method was compared to the Newton-Raphson (*NR*) and Conjugate-Gradient (*CG*) methods. The results are listed in Table 1. All three methods have step-size control and for the test cases were required to obtain essentially the same solution. All three methods were given the same starting values (initial guess) so that there was no advantage of one over the other in these respects.

Table 1 lists the number of variables (independent unknowns and dependent residuals), the number of function evaluations, and relative performance. The relative performance is the number of CPU-seconds required for the *NR* divided by the number required for the particular method (thus, *NR* will always have a relative performance of 1.0).

Test Case 1

The first test case is a nonlinear constrained curve-fitting problem. The best fitting single branch of a hyperbola was sought which would not only agree with the data (in this case experimental film boiling droplet area as a function of time), but would also have asymptotic characteristics conforming to the observed phenomena. The resulting curve fit must have one and only one root. The root must lie outside the range of the data and the derivative must be infinite at that point. The problem is nonlinear because of the constraints and the form (a rational polynomial). The partial derivatives of the residual cannot be determined analytically as these result in yet another set of simultaneous nonlinear equations. This test case was selected as being typical from among a set of 125.

Test Case 2

The second test case is similar to a nonlinear unconstrained curve-fitting problem. The values of hydraulic conductivity and storativity (groundwater analogs of electrical conductance and capacitance) were sought which would best characterize a measured field response. A field test was conducted by pumping water from a well and measuring the change in the water table in a nearby well. An analytical expression for the ideal response of an aquifer contains these two unknown parameters, which must be selected so as to best agree with the measured response. This problem is nonlinear, however the partial derivatives of the residual can be computed analytically (see note * in the table). This test case was selected as being typical from among a set of 33.

Test Case 3

The third test case is the determination of four *calibration factors* (mass transfer and pressure drop coefficients characterizing a particular type of plastic media), which are needed to run a large finite-integral code (numerical model of a cooling tower). Forty-nine sets of field data were collected for this plastic media. What were sought are the calibration factors which when input to the model would best reproduce the measured results. The finite-integral code itself was repeatedly run to provide the residuals. Needless to say, this was a very computationally intensive process—one in which minimizing the number of function evaluations was crucial. This test case, which is actually a type of inverse mass transfer problem, was selected as being typical from among a set of 6.

Test Case 4

The fourth test case is the determination of 4 phase lags and 4 corresponding weights, which would best characterize the transient response of a dammed reservoir. A linear model was sought for the cross-sectionally averaged transient flow at a specific location (adjacent to a large power plant) within a reservoir bounded by two dams, which are used for peaking (i.e., they discharge water only during times of peak electrical demand). This linear model was to become part of a larger linear systems optimization code used for long-range planning and resource management. An existing dynamic fluid flow model was used along with historical dam operations to produce a target data set. This test case was selected as being typical from among a set of 4.

Discussion

The focus of these four test cases is not on many variables, but on non-analytically differentiable residuals and lengthy function evaluations. In each case there is some physical phenomenon, which provides the basis for the residuals. Because these test cases are based on physical phenomena, the approximate bounds on the solution are also known. In each case lengthy graphical or cumbersome numerical techniques exist for finding extrema. The advantages to using the present algorithm in these cases are convenience and speed.

Table 1. Comparison of Methods for Locating Extrema

test case	variables		function evaluations			performance index		
	N	M	N-R	C-G	E-B	N-R	C-G	E-B
1	4	50	417	462	76	1	0.8	1.9
2	2	47	35*	51*	61	1	0.8	3.1
3	4	98	137	298	14	1	0.5	10.4
4	8	8760	103	344	18	1	0.4	2.4

*Note: Functions calls reporting analytical partial derivatives are more time consuming than those, which do not.

In the first test case (fitting a hyperbola with constraints and later taking its derivative) has been done for years using hand-drawn curves and a drafting protractor. The second test case (determining hydraulic conductance and capacitance) has also been done for years by graphical means and more recently by asymptotic extension to separate the coupled influence of the unknowns. The third test case (determining calibration factors for mass transfer and pressure drop) has typically been done by assuming half of the unknowns to be the same as a similar media and computing the others by *trial-and-error*.

For these test cases the average performance of the *EB* method is about 4 times the *NR* and *CG* methods. As mentioned previously, these are not isolated examples, but *working problems* from a variety of fields, which were the impetus for developing the method. The *EB* method utilizes the best features of the *NR*, *CG*, and *SD* methods along with avoiding direct calculation of the Jacobian. The relative advantage of the *EB* method was most dramatic for Test Case 3 where the difference in runtime was a matter of days (on a 33MHz-80386/7 machine).

A two-variable problem is best suited to illustrate the searching procedure graphically. Figures 1 through 3 show the contours of the function in Test Case 2 and the first few steps in the search path for the *NR*, *CG*, and *EB* methods respectively. The Z-axis or the contours is percent total residual (in 20% intervals). The dark (dense dot) region is close to the extremum and the light (sparse dot) region is far from the extremum. This graphical format was selected in order to give a *bulls-eye* appearance.

In this case the *CG* essentially follows the gradient inward to the center of the *bulls-eye* (see the dark path line in Figure 2). The *CG* path is almost perpendicular to the contours as it crosses each one. The *SD* path if it were shown would differ little from the *CG*. The *NR* and *EB* paths differ markedly from the *CG* (compare the dark path lines in Figures 1 and 3 to Figure 2). The *NR* and *EB* methods reach the vicinity of the extremum (i.e., penetrate the darkest inner contour) in significantly fewer steps than does the *CG* method.

The hybrid implementation of the present method can be seen by comparing the second step in the *NR* and *EB* paths. The line connecting the second and third points on the *NR* line (Figure 1) is almost parallel to the contour next to it (i.e., this step is almost perpendicular to the gradient). The line connecting the second and third points on the *EB* path (Figure 3) is almost perpendicular to the contour (i.e., almost in line with the gradient at the point where it crosses the inner contour). This illustrates how the *EB* method checks the search direction corresponding to all three methods (*NR*, *CG*, and *SD*) to see which is more advantageous at a particular location.

Conclusions

Broyden's derivative-free method for solving nonlinear simultaneous equations has been presented along with four enhancements. These enhancements include: a method for selecting starting values, step length control, hybrid search algorithm, and a method for escaping from extraneous entrapment. A significant performance improvement over the Newton-Raphson and Conjugate-Gradient methods is shown for four test cases taken from varied fields. Part of this performance improvement is a consequence of the derivative-free method. The hybrid search algorithm used in this enhanced Broyden method further improves the performance by utilizing the strengths of three other methods (the Newton Raphson, Conjugate-Gradient, and Steepest-Descent).

References

Broyden, C., "A New Method of Solving Nonlinear Simultaneous Equations," *Computational Journal*, Vol. 12, pp. 94-99, 1969.

Fletcher, R., *Practical Methods of Optimization*, John Wiley and Sons, New York, NY, 1987.

Morè, J. J. and D. C. Sorensen, "Newton's Method," *Studies in Numerical Analysis*, G. H. Golub, ed., The Mathematical Association of America, pp. 29-82, 1984.

Ortega, J. M. and W. C. Rheinboldt, *Iterative Solution of Nonlinear Equations in Several Variables*, Academic Press, New York, 1970.

Powell, M. J. D., "Restart Procedures for the Conjugate Gradient Method," *Mathematical Programming*, Vol. 12, pp. 241-254, 1977.

also by D. James Benton

3D Articulation: Using OpenGL, ISBN-9798596362480, Amazon, 2021 (book 3 in the 3D series).

3D Models in Motion Using OpenGL, ISBN-9798652987701, Amazon, 2020 (book 2 in the 3D series.

3D Rendering in Windows: How to display three-dimensional objects in Windows with and without OpenGL, ISBN-9781520339610, Amazon, 2016 (book 1 in the 3D series).

A Synergy of Short Stories: The whole may be greater than the sum of the parts, ISBN-9781520340319, Amazon, 2016.

Azeotropes: Behavior and Application, ISBN-9798609748997, Amazon, 2020.

bat-Elohim: Book 3 in the Little Star Trilogy, ISBN-9781686148682, Amazon, 2019.

Combined 3D Rendering Series: 3D Rendering in Windows®, 3D Models in Motion, and 3D Articulation, ISBN-9798484417032, Amazon, 2021.

Complex Variables: Practical Applications, ISBN-9781794250437, Amazon, 2019.

Compression & Encryption: Algorithms & Software, ISBN-9781081008826, Amazon, 2019.

Computational Fluid Dynamics: an Overview of Methods, ISBN-9781672393775, Amazon, 2019.

Computer Simulation of Power Systems: Programming Strategies and Practical Examples, ISBN-9781696218184, Amazon, 2019.

Contaminant Transport: A Numerical Approach, ISBN-9798461733216, Amazon, 2021.

CPUnleashed! Tapping Processor Speed, ISBN-9798421420361, Amazon, 2022.

Curve-Fitting: The Science and Art of Approximation, ISBN-9781520339542, Amazon, 2016.

Death by Tie: It was the best of ties. It was the worst of ties. It's what got him killed., ISBN-9798398745931, Amazon, 2023.

Differential Equations: Numerical Methods for Solving, ISBN-9781983004162, Amazon, 2018.

Equations of State: A Graphical Comparison, ISBN-9798843139520, Amazon, 2022.

Evaporative Cooling: The Science of Beating the Heat, ISBN-9781520913346, Amazon, 2017.

Forecasting: Extrapolation and Projection, ISBN-9798394019494, Amazon 2023.

Heat Engines: Thermodynamics, Cycles, & Performance Curves, ISBN-9798486886836, Amazon, 2021.

Heat Exchangers: Performance Prediction & Evaluation, ISBN-9781973589327, Amazon, 2017.

Heat Recovery Steam Generators: Thermal Design and Testing, ISBN-9781691029365, Amazon, 2019.

Heat Transfer: Heat Exchangers, Heat Recovery Steam Generators, & Cooling Towers, ISBN-9798487417831, Amazon, 2021.
Heat Transfer Examples: Practical Problems Solved, ISBN-9798390610763, Amazon, 2023.
The Kick-Start Murders: Visualize revenge, ISBN-9798759083375, Amazon, 2021.
Jamie2: Innocence is easily lost and cannot be restored, ISBN-9781520339375, Amazon, 2016-18.
Kyle Cooper Mysteries: Kick Start, Monte Carlo, and Waterfront Murders, ISBN-9798829365943, Amazon, 2022.
The Last Seraph: Sequel to Little Star, ISBN-9781726802253, Amazon, 2018.
Little Star: God doesn't do things the way we expect Him to. He's better than that! ISBN-9781520338903, Amazon, 2015-17.
Living Math: Seeing mathematics in every day life (and appreciating it more too), ISBN-9781520336992, Amazon, 2016.
Lost Cause: If only history could be changed..., ISBN-9781521173770, Amazon, 2017.
Mass Transfer: Diffusion & Convection, ISBN-9798702403106, Amazon, 2021.
Mill Town Destiny: The Hand of Providence brought them together to rescue the mill, the town, and each other, ISBN-9781520864679, Amazon, 2017.
Monte Carlo Murders: Who Killed Who and Why, ISBN-9798829341848, Amazon, 2022.
Monte Carlo Simulation: The Art of Random Process Characterization, ISBN-9781980577874, Amazon, 2018.
Nonlinear Equations: Numerical Methods for Solving, ISBN-9781717767318, Amazon, 2018.
Numerical Calculus: Differentiation and Integration, ISBN-9781980680901, Amazon, 2018.
Numerical Methods: Nonlinear Equations, Numerical Calculus, & Differential Equations, ISBN-9798486246845, Amazon, 2021.
Orthogonal Functions: The Many Uses of, ISBN-9781719876162, Amazon, 2018.
Overwhelming Evidence: A Pilgrimage, ISBN-9798515642211, Amazon, 2021.
Particle Tracking: Computational Strategies and Diverse Examples, ISBN-9781692512651, Amazon, 2019.
Plumes: Delineation & Transport, ISBN-9781702292771, Amazon, 2019.
Power Plant Performance Curves: for Testing and Dispatch, ISBN-9798640192698, Amazon, 2020.
Practical Linear Algebra: Principles & Software, ISBN-9798860910584, Amazon, 2023.
Props, Fans, & Pumps: Design & Performance, ISBN-9798645391195, Amazon, 2020.
Remediation: Contaminant Transport, Particle Tracking, & Plumes, ISBN-9798485651190, Amazon, 2021.
ROFL: Rolling on the Floor Laughing, ISBN-9781973300007, Amazon, 2017.

Seminole Rain: You don't choose destiny. It chooses you, ISBN-9798668502196, Amazon, 2020.
Septillionth: 1 in 10^{24}, ISBN-9798410762472, Amazon, 2022.
Software Development: Targeted Applications, ISBN-9798850653989, Amazon, 2023.
Software Recipes: Proven Tools, ISBN-9798815229556, Amazon, 2022.
Steam 2020: to 150 GPa and 6000 K, ISBN-9798634643830, Amazon, 2020.
Thermochemical Reactions: Numerical Solutions, ISBN-9781073417872, Amazon, 2019.
Thermodynamic and Transport Properties of Fluids, ISBN-9781092120845, Amazon, 2019.
Thermodynamic Cycles: Effective Modeling Strategies for Software Development, ISBN-9781070934372, Amazon, 2019.
Thermodynamics - Theory & Practice: The science of energy and power, ISBN-9781520339795, Amazon, 2016.
Version-Independent Programming: Code Development Guidelines for the Windows® Operating System, ISBN-9781520339146, Amazon, 2016.
The Waterfront Murders: As you sow, so shall you reap, ISBN-9798611314500, Amazon, 2020.
Weather Data: Where To Get It and How To Process It, ISBN-9798868037894, Amazon, 2023.

www.ingramcontent.com/pod-product-compliance
Lightning Source LLC
Chambersburg PA
CBHW070243220526
45465CB00004B/1504